Jobs, Robots & Us

Jobs, Robots & Us

Why the Future of Work in New Zealand Is in Our Hands

Kinley Salmon

BRIDGET WILLIAMS BOOKS

First published in 2019 by Bridget Williams Books, PO Box 12474, Wellington 6144, New Zealand

ISBN 9781988545882 (Paperback), ISBN 9781988545851 (EPUB)
ISBN 9781988545868 (Kindle), ISBN 9781988545875 (PDF)
DOI https://doi.org/10.7810/9781988545882

This book is written in a personal capacity and reflects only Kinley Salmon's personal views.

Acknowledgements
The publishers would like to acknowledge the ongoing support of the Bridget Williams Books Publishing Trust, which enables all BWB publications. Particular acknowledgement is made to the Alterno Foundation for their generous contribution to this publication.

Alterno
FOUNDATION

A catalogue record for this book is available from the National Library of New Zealand. Kei te pātengi raraunga o Te Puna Mātauranga o Aotearoa te whakarārangi o tēnei pukapuka.

Edited by Caren Wilton
Internal design and typesetting by Tina Delceg
Cover design by Jo Bailey
Author photo by Erin Salmon
Printed by Printlink, Wellington

Contents

Introduction
Getting a grip on the future of work

Harry and I believed that in the interests of economics, if nothing else, mechanical aids should be used anywhere where they did not affect either the aesthetics of the end product, or degrade the life of the worker. Machines are like that. At first they reduce the monotony and enhance the life but in the end, if used indiscriminately, they increase the monotony and impoverish the life of both maker and user.

– May Davis, my grandmother, *May Davis: Her Story*

A tall, handsome man in a slim-fit blue suit, open-collared white shirt and tan shoes bestrode the stage. His luxuriant hair was swept back and as he spoke he flashed a winning smile. He had charisma. He had a vision. Above all he had confidence. He wasn't here just to talk about the future; he was here to embody the future.

Everything was moving exponentially, according to this messenger of (from?) the future. Nothing was too sturdy to be disrupted. Nothing was too bold to merit a PowerPoint slide. His other talks had made similar claims. Manufacturing off-planet? The next big business opportunity. Driving a car? Practically a thing of the past. Taking a family trip to space with his two kids? A certainty. Banks? Soon won't be needed thanks to blockchain. Living to the age of 120? The new 60.

Rob Nail, the chief executive officer of Singularity University, certainly knew how to make an impression. The audience watched on with admiration, fear and bewilderment. From his good looks and persuasive sheen of success to the way he conjured the extraordinary future awaiting us if we just made the right choices, Nail brought to mind another character of American life: the televangelist preacher. All that was missing was someone standing up in the back row and shouting 'Hallelujah!'

Indeed, like a preacher, belief was his constant refrain. Belief in the supremacy of technology. Belief in the power of entrepreneurs. Belief in fundamental disruption. When challenged, he always fell back on his very strong belief. That may be normal in a church. It is less common from the head of a university.

By the end some in the crowd were put off by the sheer bravado of it all. Yet most of the companies Nail talked about were real. Someone was funding them. Someone believed enough to stump up the money.

I left the cavernous auditorium perplexed and disconcerted. Human technological progress has already been extraordinary. We can fly around the world, cure devastating diseases and chat instantly with loved ones across the oceans in high-definition video. Singularity University has had some powerful backers, including Google and the famous futurist Ray Kurzweil. Some of what Nail said had to be correct. Yet it was implausible that it all was.[1] But how much was accurate? And which parts? And what does it all mean for we humans who have built modern society around the idea of getting up in the morning and going to work?

Nail's talks – many examples of his slick work are available on YouTube – capture neatly the current debate around technology and the future of work; there is so much hype that it is hard to think straight. Often the stories we tell and are told jump in a nanosecond from optimistic press releases to a world of mass unemployment with robots doing everything imaginable. The way technological changes are reported often doesn't help. As David Brougham, a business professor at Massey University, told me, 'You have to fit with the media's narrative of fear to get your results out there.'[2]

Still, there is plenty of technological change to see, even without relying on the media. Before I began writing this book, I'd seen a robot mowing the lawns, walked past another robot doing food delivery, and eaten a vegetarian burger with an incredible likeness to meat. This was in the United States and Europe. In New Zealand we've seen a drone deliver pizza, driverless shuttles are used at Christchurch airport, and Google suggests snappy responses to our emails (albeit in exclamation-mark-heavy Silicon Valley style).

Many people welcome these new innovations with excitement. Technological progress, after all, has already changed human life dramatically and extraordinarily for the better. Further breakthroughs could do tremendous good. Writing for *The Economist* in 2016, I noted how algorithms could help identify children at risk of lead poisoning in houses in Chicago and get them out before they were affected. In the same article I discussed how algorithms could dramatically improve our ability to predict heart attacks in hospitals. More recently, there is research showing how skin cancer – a scourge in New Zealand – could be detected far more easily using artificial intelligence (AI).[3]

In the workplace, better technology can also make work dramatically safer and more humane.[4] Technology means we can do things that were previously unimaginable. Imagine telling Captain James Cook or Hōne Heke Pōkai, the first Māori signatory of the Treaty of Waitangi, that people fly around New Zealand all the time and that dying of measles or tuberculosis is now rare here.

More fundamentally, technological progress is the main reason we are better off than our ancestors. Better technology means we can produce more with the same effort. Technological progress raises productivity, the key measure of the economic output we produce per hour.[5] Higher productivity makes us wealthier and sits behind the entire history of modern economic growth. A more productive New Zealand would be a richer New Zealand.

Yet that sunny paragraph sits oddly with much of what we hear about technology and artificial intelligence today. Indeed, new technologies can also be very scary. Artificial intelligence of the kind that uses vast reams of data about everything we do – big data with echoes of Big Brother – raises major privacy concerns.[6] Many elections have been swamped with fake news and psychologically targeted advertisements thanks to algorithmic analyses of our social media activity. All this deserves attention. The focus of this book, however, is on a perhaps even bigger area of concern about technological change: the future of work.

Being deeply worried about what may happen to work is understandable. If you are a radiologist hearing about how artificial intelligence can spot cancer, despite a certain professional positivity, it is hard not to focus on the looming spectre of unemployment. For taxi and truck drivers every news article on driverless vehicles casts a shadow of uncertainty on the future. Even hotel workers are growing concerned about automatic check-in systems threatening their job security. These concerns have been met with renewed enthusiasm for the idea of a universal basic income, to deal with widespread unemployment, and with it the implication of a radically different economy and society to the one we currently live in. The idea of losing your job to a robot is indeed deeply alarming. Some studies have suggested nearly half of New Zealand's jobs could be eliminated and taken over by machines by the mid 2030s. That's scary. It's also, as this book will argue, misleading. Change is coming, as it always does, but not like that. Moreover, we are not powerless to shape the future.

To be clear, today there is a pretty large flaw in the technological unemployment story; almost every economic indicator that could suggest impending technological unemployment (unemployment due to new technologies) instead shows the opposite. More working-age New Zealanders are in work than ever before in history.[7] Unemployment is low. Labour-force participation (the share of people who are working or actively looking for work) is high and rising. Productivity growth, perhaps the key measure of widespread automation in the economy, is low and has been stubbornly (and problematically) low for more than a decade. Moreover, we have seen dramatic technological breakthroughs before: the steam engine, electricity, the phone, and the computer. Yet despite these inventions, here we are still with more people working than ever.

These cold hard economic measures are hard to reconcile with the technological change we hear about and, to a lesser extent, see. The hype can be deafening. Public figures talk frequently about travelling to Mars. There is widespread speculation about immortality. Vivek Wadhwa, a *Foreign Policy* magazine Top 100 Global Thinker and Distinguished Fellow at Carnegie Mellon's school of engineering, has claimed that 80 to 90 per cent of American jobs will be eliminated in the next ten to fifteen years.[8] No one wants to miss out on the artificial-intelligence excitement. In the last three months of 2017 public companies mentioned 'artificial intelligence' or 'machine learning' over 700 times on their earnings calls. That is seven times more than in the last three months of 2015.[9] This book aims to hit pause on the hype long enough to unpack what we know, what we don't know, and what might really be going on.

This book also hopes to help start a debate around what kind of future we should aim for by reminding us of our ability to shape the future – and grounding the arguments in evidence. We can, after all, set our own criteria for how we use and engage with machines. That's exactly what my grandparents, potters Harry and May Davis, did. Yet as May suggests in her book, quoted above, there can be a trade-off between economic maximisation and human meaning. Addressing this trade-off and deciding how we engage with machines ultimately requires value judgements about how we should balance different goals like wealth, meaningful work, freedom, leisure, environmental protection and social connection. At the core of these judgements is one big question: what does the good life look like? In setting out the trade-offs and challenges that may come with two broad possible futures for New Zealand, this book aims to help the reader find their own answer to that question.

The future has long fascinated me. I first studied 'the future' in my English class as a sixteen-year-old student at Nayland College in Nelson. I remember being struck by how dystopian visions far outweighed even mildly positive ones. George Orwell's *1984* is not exactly joyful. Nor was the movie *Gattaca*, in which babies are genetically engineered, leaving Ethan Hawke's character, conceived in the lusty traditional way, a victim of discrimination and needing to commit genetic fraud to follow his dream of being an astronaut.

I am not a wild optimist, yet it strikes me that, despite what we often hear, the future could well be better than the past. Indeed, the evidence suggests that this has historically been the case. As Princeton professor Angus Deaton recently showed in his academic blockbuster *The Great Escape*, most things are actually getting better.[10] Almost everywhere, and certainly in New Zealand, we are generally richer, live longer and are more educated than forty years ago. Of course, averages can hide challenges for individual groups. But we should not fall into the habit of simply assuming the future will be worse.

There are three other problems with popular depictions of the future that this book sets out to challenge. These problems are as common today as when I first noticed them back at Nayland College. First, in most representations, the future just arrives fully formed: it is not something that we create. Stories have to start somewhere, and that may well be in the midst of a dramatic dystopian future, if one is writing fiction. Yet in reality as time progresses humans make countless choices to shape the world they live in. There is a reason we say history is something we 'make'. When it comes to technology, we shape the kinds of technologies that are invented and we choose what we do with the new capabilities these technologies provide. We should not forget our own agency to shape the future.

Second, most stories about the future are not about New Zealand's future specifically. The academic studies, news stories and narratives driving debates about the future of work are often from overseas – most commonly from the United States of America and the United Kingdom. That's understandable; there is a lot to be learnt from experiences elsewhere and this book will highlight many, but it should not be done unthinkingly. Not everything that happens overseas applies in the same way in New Zealand. We should remember to pause and ask how and whether this story, study or idea really fits here.

Third, while it is my generation's future at stake, it is largely not my generation writing about it. Few thinkers, leaders and writers on this topic are millennials. It will be young Kiwis like me who will be around in this future; we will be alive but (we are told) unemployed. I wrote this book because it is my future, and the future of young New Zealanders like me, that I keep hearing about. Will it really change so fast? Will it really take shape regardless of what we want? Will it really be that bad?

As Danish physicist Niels Bohr quipped, 'Prediction is very difficult, especially about the future.' While this book is not narrowly an effort at prediction, writing about the future always involves considerable uncertainty and leaves ample room for disagreement. Mine is only one perspective. It is certainly not intended as an attempt at the last word. I hope, above all, that it encourages other perspectives to be heard and debated. In navigating the uncertainty, I have been extremely fortunate to draw upon interviews and discussions with a wide range of academics, policymakers and private-sector leaders both in New Zealand and overseas.

The book also draws extensively on a wide body of published research.[11] It is grounded in economics but includes many other disciplines and perspectives. History shows how technological revolutions have affected work before. Psychology sheds light on how different possible futures might actually leave people feeling. Politics and sociology highlight attitudes to change and the scope of political choices we may have. This book also puts environmental analysis squarely in the discussion on the future of work. It belongs there because a world of far higher productivity and income is also likely to be a world of far higher consumption. So far in human history far higher consumption means far greater environmental damage and pollution. Yet, today more than ever, we cannot afford to ignore the environment.

Before we go any further, what do we actually mean by artificial intelligence, automation and robots? Artificial intelligence is not a machine that can reason and think like a human. We are very far from machines with true human-like intelligence. Instead artificial intelligence, which is sometimes called machine learning, is basically very powerful computers churning through vast reams of data to find patterns and make predictions. Think of artificial intelligence as akin to a crack team of statisticians – but faster.[12] Using a technique called deep learning, these computers can identify patterns and make predictions without being explicitly programmed to find a particular pattern. Artificial intelligence is excellent at prediction and pattern recognition – identifying thousands of pictures of cats, recognising red stop lights, and spotting irregular financial transactions – but struggles at rudimentary tasks like basic reasoning. New situations tend to baffle artificial intelligence – if there is no pattern it gets stuck.[13] Nonetheless, this predictive and pattern-recognition ability is what powers things like computer translations, targeted advertisements and driverless cars. Given increases in computing power there are many ways artificial intelligence could be applied, particularly as the data these computers can process includes not just text and numbers but images and sounds too.

Artificial intelligence is only one of the technological advances people are talking about. Automation is another word frequently bandied about. Automation can be powered by artificial intelligence but does not have to be; 3D printers can automate some manufacturing but do not require artificial intelligence. In this book, I will use 'automation' to refer to any kind of technological change, whether artificial intelligence or not, that directly eliminates a worker from a specific task. For simplicity, I will also use 'robots' as shorthand for both artificial intelligence and automation. Yet there are also other kinds of technologies and innovations that are not either explicitly automation or artificial intelligence; the invention of the mobile phone or electric car, for example. This book will underline how important it is to recognise that not all technological progress and innovation is automation or artificial intelligence.

At this point, let me at least attempt to predict a simpler future: the content of the subsequent chapters.

To grapple effectively with the question of whether technological change is soon going to leave us unemployed we need to understand where new technologies come from, how they arrive in our workplaces and homes, and what influence we have over both these processes.

This book therefore begins in Chapter One with two simple but rarely discussed questions: why do we get some kinds of technological innovation and not others? And can we influence the technological innovation we get? Despite the way it is often talked about, technological change does not arrive from outer space. Instead we humans have rather a lot to do with the kind of technology that gets invented – influencing the direction technology takes is not beyond us.

Chapter Two then homes in on the second big under-discussed issue: technological diffusion and adoption. How is it that new technologies make it into our lives and change how we live and work? It highlights the absurdity of assuming, as too many studies do, rapid widespread adoption of every technology that has been mooted or demonstrated in a research laboratory. For robots or artificial intelligence to have a serious impact on work, they do not just have to be invented; they must be adopted across New Zealand. Yet New Zealand's record at adopting new technologies at work is weak. This chapter discusses why that is, whether the speed of technological diffusion and adoption is accelerating, and what is likely to determine how quickly new technologies are adopted in the future.

Understanding the reality of technological innovation and adoption can help us see clearly amid the alarming predictions of job losses. Chapter Three is therefore the time to tackle the big question: are the robots really at the gate with designs on our jobs? If that were the case we would expect to see rising unemployment, declining labour-force participation, rapid productivity growth and a lot of people being forced to change occupations. Yet in New Zealand, and in most rich countries, we see exactly the opposite. In fact, as I write in late 2018, we see historically noteworthy opposites: record low unemployment in much of the world, record high labour-force participation, record low productivity growth and record low occupational churn. The chapter also explores whether the few exceptions to these figures might for some reason be a better guide to the future and whether big changes could be just around the corner as new technologies come on stream. While major changes to how we work may be ahead, they are unlikely to be historically unprecedented in New Zealand and may be a smaller upheaval than in other countries.

Predicting the future of individual technologies is a famously difficult (some might say foolish) task. Chapter Four therefore zooms out and considers what may happen to employment and wages *even if* we do see rapid technological progress and considerable automation and use of artificial intelligence in New Zealand. To help think this through it sets up a simple model grounded in economics. Automation and artificial intelligence do displace workers but, as this model shows, that is not all they do. They also increase the demand for workers and help to incentivise the creation of new tasks for human workers to do. These effects are what have kept employment high despite extraordinary productivity growth and automation in the past two hundred years of human history. Could the historic balance between job displacement and creation be breaking down and, if so, why? Fears abound of good middle-class jobs being hollowed out by technological change, wage stagnation and insecure work. While not all of these apply equally to New Zealand, these are reasons for concern. But they are not an inevitable result of technological change. They are also a reflection of our political choices.

Chapters Five and Six then sketch out two very different possible futures for New Zealand at least twenty-five years into the future in which technology has eventually progressed significantly. In Chapter Five I consider a world in which we continue the pattern of modern economic history: a society structured around most people getting up in the morning and going to work. Grounded in the evidence already outlined in the book, the chapter begins with a depiction of life for an imagined Wellington couple, Paul and Natia. It discusses why this vision is plausible, even in the face of considerable technological change. Yet to reach Paul and Natia's version of New Zealand, the chapter argues, we would need to overcome six large challenges.

The following chapter presents a very different world, one in which having a paid job is an exception rather than a norm. Rachel and Wei, inhabitants of a future Nelson, enjoy the fruits of technology thanks to a gift of shares in technology companies from the state and a basic income. This kind of world is more and more commonly discussed today and this chapter takes its implications seriously – for everything from education to economic growth. Yet ending up in such a world soon is not technologically inevitable, despite what we sometimes hear. To get there would require intent – and acceptance of some real trade-offs. The chapter highlights three fundamental challenges that we would need to overcome if we wanted to reach it.

Chapter Seven is about choice and our ability to shape the future in the long run. It sets out how, to a significant degree, which of these futures we head towards in the long term is up to us. Where we end up has far less to do with any law of economics or inevitable technological change and far more to do with the choices we make as a country. Establishing an overall direction for New Zealand should be the result of a robust democratic process – especially as there are trade-offs for any choice we make and difficult policies to pursue no matter the kind of future we might favour.

Indeed, many of the policies discussed in this book are politically significant and challenging. The political difficulty of change is rarely forgotten. This book acknowledges that, but also complements it by underlining the fact that over time policy, business choices and political changes can be extraordinarily influential and epoch-shaping in ways we tend to forget. Moreover, by recognising the enormous importance in the long run of our choices and policies, we may prove more able to make these politically challenging choices. This recognition can also help us focus on policy areas we might otherwise ignore as irrelevant to the future of work.

If we do not see how significantly our political and policy choices shape the future, we risk inadequate political engagement. That could leave us sleepwalking into a worst-of-both-worlds future that suits very few of us. Our existing policies and choices are, after all, already shaping the future. As such, recognising how much we can shape the future of work is the first step to engaging in the political and public debate to do so. That may help bring things that currently seem politically unattainable into the realm of the possible.

Finally, we do have some time to make these choices. This book presents long-term visions of the future and, as we will see, technology is not moving quite as swiftly as is often suggested. Policy change, however, does not happen overnight either. Now is therefore the moment to start debating and choosing what kind of future we want to head towards.

The book concludes with my own view of which future is worth pursuing. I will leave this for the reader to discover. Overall, this book aims to do three things: to help outline realistically where we stand today on the question of employment and technology; to provide a way to think about how changing technology may impact work and employment as and when it does occur; and, above all, to help us reclaim our own sense of agency to shape the future, and in doing so to encourage us to begin today the debate over the future we want for New Zealand.

And so, to the future.

One
Of innovation and influence

Too often technology is discussed as if it has come from another planet and has just arrived on Earth.
– Sir Anthony Atkinson, quoted in Rotman, *MIT Technology Review*

'Rise of the machines: How soon until robots take your job?', 'The robots are coming – is your job safe?', 'Robots on the march into retirement homes', 'March of the fembots: why robots are designed to be female', 'Invasion of the robot workers', 'Countries best prepared for robot revolution', 'Surviving the rise of AI machines', 'When the robots are smarter than us', 'Rise of the machines: robots set to take over in bed', 'Robots threaten millions of jobs', 'The robot that wants to start a family'.[1]

These headlines from major New Zealand news websites have robots taking your job, coming, marching, invading, threatening, rising, and even taking over in bed. Two things leap out from these headlines. First, we're surely doomed – more on that later. Second, the robots themselves are the ones *doing* these things. They are the ones taking, marching, rising, threatening (and wanting to start a family!) They already sound human. They certainly sound like they have agency and plans – and a peculiar love of marching as their preferred form of transportation.

Even when the media notes that there is doubt over whether the technologies will actually doom humans, it tends to be discussed as if the development of technology is something we have no control over – nothing more than a game of dice. Get a six and we are doomed, get a five and some of us at least may still have a job. Human choices and actions are strangely absent. 'Human engineers design robots that may take some of our jobs' doesn't have the same ring. Nor does 'Technology companies march robots towards us'. A little journalistic flourish is understandable. But these kinds of headlines belie a central fact about technological change both today and throughout human history: we humans are making these technologies.

What is happening in technology, innovation and the future of work is not an inexorable process beyond anyone's control. Instead it is the result of choices by politicians, business people and consumers. These choices shape the innovation environment and have profound impacts on what kind of innovation occurs and what impact it has on the world of work. To make this case, the next two chapters will tackle two rarely asked questions. First, why do people and businesses invent the kind of things they do? Second, how quickly do, and might, these inventions spread around the world and change our working lives?

Answering these questions will help interpret the flurry of stories about the 'jobpocalypse' and the future of work. It will suggest that New Zealand is far from powerless in the face of a technological tsunami. Much cutting-edge innovation happens abroad, but New Zealand also has its own innovation economy, which the government and business can consciously shape.

New Zealand also has the chance to shape the debate on technological innovation worldwide as a first mover and an intellectual leader to inspire others – a role we have played before in different arenas, from women's voting rights to inflation-targeting monetary policy. Our government's policy choices also strongly influence the extent to which technologies developed abroad are adopted in New Zealand and impact, or fail to impact, the workplace. There is so much we as New Zealanders and our government can do, once we decide what it is we want from the future.

This chapter, and this book, will attempt to move beyond seeing new technologies as simply good or bad. While new technologies are very often positive, they are not inevitably and always so. How we should view a given innovation depends on what kind of innovation it is, what effect it has, and whether it actually addresses a real societal need. These are all things that government, business and consumer decisions can shape. Above all, the value of new technologies depends on the kind of society we are aiming for.

So why do we get the kind of innovations that we get?

Innovation comes in more than one flavour

Economists have been oddly slow to acknowledge the influence of policy on technological innovation. In 1956 Robert Solow and, separately, the Australian economist Trevor Swan introduced what became known as the Solow–Swan growth model. This claimed that it was technological innovation that truly determined long-run economic growth. Most economists, and the Nobel Prize committee, agreed.[2] Yet in this model, and for a long time, economists largely thought about technological innovation as 'exogenous' – a bolt from the blue, the genius of a cranky inventor in their workshop, a eureka moment. These eureka moments were assumed to be unaffected by the world around them.

This view began to change when in 1986 Paul Romer (later to be winner of the Nobel Memorial Prize in Economics in 2018) introduced a model that viewed technological innovation as 'endogenous' – determined by the economic system itself.[3] Education, knowledge and research were now seen as critical to generating technological innovation and economic growth. Innovation policy suddenly really mattered. Gordon Brown, the former Chancellor of the British Exchequer and later Prime Minister, was a fan. He famously declared that the British Labour Party's approach was 'rooted in ideas which stress the importance of macro-economics, post neo-classical endogenous growth theory and the symbiotic relationships between growth and investment, and people and infrastructure'.[4]

Brown was roundly mocked for his jargon explosion but to many economists he made sense. Endogenous growth theory is widely taught in university economics courses. How to generate innovation and growth is an overriding preoccupation among governments around the world. Yet when it comes to linking innovation and technology to the future of work we too often forget about our own agency in shaping what happens. That is a mistake. When asked for the most important lesson from his work, Romer said, 'What happens with technology is within our control.' He went on to add that 'if we set our minds to improving technology we can improve it in a direction that seems to be important to us and at a faster rate ... Instead of treating it like the weather, we can treat it as something we control.'[5]

Innovation is a word so often used that, like overcooked vegetables, it has lost all shape and flavour. Getting more specific would help. There is not one single type of innovation; it is useful to think of innovation as coming in different varieties. Harvard Business School professors Clay Christensen and Derek van Bever set out one model for thinking about types of innovation from the perspective of an innovation's impact on jobs and growth.[6] They identify three categories: efficiency, sustaining and market-creating innovations.[7]

Why bother with such distinctions? Because when it comes to their direct impact on jobs, these three types of innovation differ greatly. Government efforts to generate innovation do not always produce the same type. Similarly, some sectors are more prone to certain types of innovation than others.

Efficiency innovations enable businesses to make and sell established products for lower prices. This kind of innovation often involves process improvements or adjusted business models. Stephen Tindall's Warehouse was a classic efficiency innovation in retail – using scale to deliver the same products for lower prices. Efficiency innovation does two things. It increases productivity, what we can produce in a given hour, which often has the side effect of reducing the number of workers needed to produce the same amount of stuff. Second, it can free up money to use on other things by, for example, reducing the amount of money that is tied up in holding stock in a shop. The kinds of innovations we hear about in the debates on automation are often efficiency innovations. For example, bots that take over tasks like answering customer queries online or in call centres mean the same service can be provided at a lower cost.

Sustaining innovations improve existing products. Consider the iPhone 7 and iPhone 8. The innovations in an iPhone 8 are pretty cool, as any enthusiastic Apple customer can (and, I find, will) tell you. But even the most excited Apple customer queueing outside the sleek silver store isn't likely to buy an iPhone 8 and keep using the iPhone 7. They're going to upgrade and stop using the iPhone 7. Similarly a new customer buying a phone for the first time will not buy both the iPhone 7 and the iPhone 8. Sustaining innovations keep the wheel of innovation and growth moving, but don't tend to generate large numbers of jobs – precisely because the new product substitutes for the old one. The jobs and people that used to support the development, sale and apps of the iPhone 7 now support the same for the iPhone 8. Apple does not hire an entire new set of workers, sales teams and app developers for the iPhone 8. Instead the workers who did that for the iPhone 7 gradually shift their focus to the iPhone 8.

Market-creating innovations do what they say on the tin; they generate new markets. Specifically, they create a market among 'non-consumers' – people who do not currently buy a product. This can happen in two ways. First, by inventing something that people want to buy that did not previously exist. Second, by creating a radical new business model that makes an expensive niche product affordable to whole new groups of people. These two steps often happen sequentially, as a new product can initially be too costly for everyone except the rich, but radical new business models can bring it to large new markets. Moreover, just as market-creating innovations can get people who did not previously consume something to buy it, they also allow people who were not previously employed to be employed in order to serve these new customers. Market-creating innovations therefore help create jobs.

There are many examples of market-creating innovation.[8] Mainframe computers cost hundreds of thousands of dollars and very few people used them.[9] The innovations that led to the personal computer were market-creating innovations that made computing power available to hundreds of millions of people. This generated an utterly extraordinary range of new jobs in computers and information technology, from manufacturing to digital content to sales and programming. One estimate suggests the introduction of the personal computer enabled the creation of 15.8 million new jobs in the USA, even after accounting for jobs displaced.[10]

Similarly, thanks to innovative simple design and the use of assembly lines, the Ford Model T was the first low-cost automobile. It allowed millions of Americans to buy a car for the first time. Mass automobile usage created an explosion in employment in automobile-related industries: car production, repairs, mechanics, tourism and fuelling. Market-creating innovation can happen in agriculture too. In Chile innovations in logistics and cold storage as well as cultivation meant farmers could sell out-of-season fresh fruit and vegetables in North America during winter (again targeting non-consumption) and increase employment to do so. The thing to remember about market-creating innovation is simple: it is the most likely to create jobs.[11]

Distinguishing between types of innovation with different impacts is increasingly common. Christensen and van Bever's distinctions are not the only ones on offer and their distinctions are not perfect; not every innovation maps neatly into them.[12] One of the world's most cited economists, Daron Acemoglu of the Massachusetts Institute of Technology (MIT), prefers to distinguish between enabling and replacing technologies.[13] Enabling technologies help drive growth and jobs by expanding what humans are capable of, while replacing technologies take over tasks from humans entirely – much like efficiency innovation can. Yet there is broad consensus that there are different kinds of innovation with different effects on jobs, and that the mix of these different kinds of innovation will be hugely influential on what happens to employment in the future.

Empirical studies also support the idea that some kinds of innovation are more likely to help create jobs.[14] Innovation that creates a new product, the most obvious form of market-creating innovation, has the strongest positive job-creating effects. This is particularly pronounced in higher technology sectors, perhaps because the leap of innovation in the new product is greatest. Similarly, spending on research and development (R&D) by firms is also associated with stronger employment growth in higher-tech sectors.[15] This may indicate that substantial R&D investment is needed to create a product that goes beyond simply sustaining innovation, like the next iteration of the iPhone, into truly market-creating innovation, like inventing the mobile phone in the first place. Empirical work backs up the other side of the theory too: innovation that focuses primarily on process, one common form of efficiency innovation, often reduces employment.[16]

Silicon Valley is abuzz with innovation. A fair chunk of that buzz is about a very particular kind of innovation: better connecting owners of a product with buyers of that product through a mobile application. For example, the app Wag!, which was valued at US$650 million in 2018, connects people who want someone to walk their dog with someone trustworthy and willing to walk dogs for money. New Zealand has its own mobile app platforms like Zoomy, a ride-hailing app, and Parkable, a parking app.[17]

By connecting buyers and sellers, these apps are doing what a website like TradeMe does, or before that the classified section of the local newspaper. However, they are doing it more effectively, using Global Positioning System (GPS) data, and at a larger scale. The most successful versions of these generate value by allowing people to better use an existing asset like a room, car or parking spot. That is the basic premise behind Airbnb, Zoomy or Parkable. Kate has a spare room, and would like to rent it out, Tui is visiting Wellington and needs somewhere to stay – bingo!

While these apps are often pitched as disruptive, market-creating innovations, many of them actually look much more like sustaining innovations. The core product (a room to stay in, ride somewhere, parking spot or person to walk your dog) is not new, although it is improved in terms of convenience and experience. Sustaining innovations of this sort also have the tendency to come at the expense of existing providers of the same core product, such as hotels, regular taxis and parking buildings. Just as the iPhone 8 substitutes for the iPhone 7, the Airbnb room substitutes for the hotel room. The furore over Airbnb and Uber worldwide is a testament to this.[18]

If such innovations were primarily market-creating, they would generate consumer spending that did not directly substitute for existing spending on hotels and taxis. In Christensen's terms, they would target 'non-consumption'. Thanks to the improved marketplaces for connecting buyers and potential sellers these apps do encourage some new consumption. Uber, for example, has made personal car services cheaper, and for this reason some would argue it has been a market-creating innovation. Yet it has been able to do this largely by heavily subsidising the price of a ride using money from investors. Uber lost US$4.5 billion in 2017 in part by doing exactly this.[19] Its sustainability remains in question.[20] As *The Economist* put it, 'firms that burn up $1bn a year are sexy but statistically doomed'.[21] Uber may not be doomed, but its ability to be a market-creating innovator without any other innovations may be.

To recap, we have efficiency innovation, which reduces the capital and labour needed to produce a given product. This can free up money and people to invest in and work on other things, perhaps including generating other kinds of innovations. Sustaining innovation updates products and maintains the existing number of jobs. Finally, market-creating innovation creates new markets and generates new jobs. All of these types of innovation have an important role. The risk if automation dominates technological progress is that we get a lot of efficiency innovation, which reduces jobs, and not much market-creating innovation, which creates them. If the capital freed up through efficiency innovation isn't invested much in market-creating innovations, things can go badly wrong. Understanding these three types of innovation can help us see more clearly what is happening amid the technological hype today.

The mother of invention isn't who you think she is

Why do we have such successful drugs for erectile dysfunction? People are willing to pay a heck of a lot to get an erection. Why do we have drugs for male pattern baldness? The same logic: drug companies rake it in for keeping hair on heads.[22] Why don't we have effective vaccines for malaria (445,000 deaths in 2016) or HIV/AIDS (940,000 AIDS-related deaths in 2017)?[23] One big reason they have not been invented is because it has not been especially profitable to do so.[24]

It's hard to escape the relationship between expected profit and innovation. Even where unprofitable medications have been invented for particular diseases, profit is often still at the heart of the R&D rationale. In central Africa, being bitten by blackfly can have dire consequences. Thousands of fly larvae circulate in your bloodstream and gather together under your skin and in your eyes. Excruciating itching, lesions and blindness follow. Yet thanks to the drug Ivermectin these larvae can be killed before they blind you. Merck, the pharmaceutical company, now gives this drug away for free in affected countries in Africa, and has been lauded for doing so. Is this a heart-warming tale in which research and innovation was driven by the desire to cure a horrifying disease? Not quite.

This drug was originally developed to treat parasites in animals and was launched on the market in 1981 as Ivomec. It quickly became the largest-selling animal health product in the world with annual sales of around US$1 billion. By 1987 it was Merck's second-largest selling product overall. Merck deserves great credit for donating the drug for humans once it was developed – although analysts have suggested that after tax deductions and public image benefits the donation may be a net financial gain to Merck.[25] Yet, and this is the key lesson for the way innovation works, Ivermectin would probably not have been invented if it weren't for the highly profitable animal market. As one study put it, the drug's effectiveness for animals allowed Merck to 'reap large profits on these innovations from the animal health market, which justified the fixed costs of research and development'.[26]

The proverb 'necessity is the mother of invention' may be true when we've lost the tent pegs on a camping holiday, but in the world of privately funded R&D 'expected profit is the mother of invention' is less snappy but more accurate.[27] There is of course great uncertainty and serendipity inherent in the process of R&D. But in the long run the kinds of innovations that get the most effort, investment and opportunities to succeed in the private sector are those that are expected to be most profitable.

Profit's influence on technological innovation is widespread. Take the debate on how to combat climate change. Savvy activists want governments to put a higher price on carbon – in other words to require companies to buy an expensive right to emit carbon. Why? If emitting carbon became an expensive part of production, companies would be willing to pay for technologies that reduce carbon emissions. The incentive to invest in green innovation would suddenly be much stronger. While this logic is widely accepted in the debate around innovation to combat climate change, the same holds, but is less often considered, when analysing innovation more broadly. We get labour-saving innovation because in the markets where innovation occurs, labour is expensive relative to other business costs.[28] Let me explain.

Innovators want to make money and so do businesses that use new technologies. Consider Kentucky Fried Chicken's restaurants in New Zealand. To produce your chicken burger KFC needs bread, chicken, salad, premises, power and water, and workers to whom they pay wages. Economists call these inputs, and the prices of inputs determine a business's costs.

One way for KFC to increase profits is to reduce costs. In many rich countries, wages are a sizeable part of the cost of your lunch. In New Zealand restaurants and cafés, paying workers can cost about a third of revenues.[29] As a result, innovations that reduce the need to hire staff, labour-saving innovations in the economics lingo, are often bought and deployed with gusto. KFC's Auckland city branch has done just this by installing kiosks to replace humans who previously took orders. This choice rewards the innovators who develop labour-saving products like kiosks. It also encourages other innovators to find new ways of reducing the need for staff to reduce labour costs. Over time, these deep economic incentives really matter.

It is not only the absolute price of inputs that matters, it is also how much they cost compared to other inputs: their relative price. When wages are expensive relative to other inputs, the innovation system tends to focus most on innovations that reduce the need for workers. Businesses after all want to conserve on the most expensive input to their production as a priority. However, if bread were extremely pricey, KFC would pay for innovation that might help reduce that cost (synthetic bread perhaps?). It's hard to escape this logic. As Acemoglu told me, 'What automation does is about finding the cheaper factor for producing a good', and for that reason 'factor prices become more important'.[30] It really matters which input is most expensive.

If the prices of things companies use in production – inputs – are so important to the kind of innovation we get, we really want to make sure that the price of these inputs makes sense. Prices for things in demand should reflect scarcity. If there is hardly any of something it should be expensive. If there is a lot, it should be cheaper. Yet markets can fail, or simply not exist, meaning that prices do not reflect scarcity. The carbon-absorption capacity of Earth's atmosphere is a scarce resource but there has historically been no market at all for this scarce resource – it was just a free-for-all. That can leave the innovation system focusing on the wrong thing. As Nobel Prize-winning economist Joseph Stiglitz puts it:

> [T]he real scarcity in the world today is related to our planetary boundaries: if we 'ruin' this planet through an excessive emission of greenhouse gases, we cannot move to another. This means that from a social point of view we should be especially focusing on innovations that reduce emissions; but so

> far, without a carbon price, firms have little incentive to do so. At the same time, we see firms devising innovations leading to more unskilled labor unemployment.[31]

There is a price on carbon in some countries, but it is far too low. Innovators and businesses alike therefore focus too much on saving on labour rather than carbon, not necessarily because labour is too expensive but because carbon emissions are too cheap – and incorrectly so.[32] This is economically inefficient; we are innovating most to save on the wrong thing. Moreover, carbon isn't the only input to production whose price doesn't reflect its true scarcity. The same is true of other resources such as water in much of New Zealand.

Workers may well be one of the hidden losers from the world's failure to fix market failures in things like carbon and water. The results of these failures and skewed incentives are at times evident in the very same industry. In aviation, we already have self-check-ins and progress toward true autopilot planes, both of which reduce labour costs for air travel. Yet we are a long way from what is really needed: zero-emissions commercial flights.

There is a real trade-off here. In the short to medium term there are only so many researchers and innovators. This means, as Acemoglu and his co-author Pascual Restrepo put it, that 'there is a natural trade-off between faster automation and investments in other types of technologies'.[33] Put differently, what impact could the world's best engineers, scientists, and investors have if they were properly incentivised and rewarded for economising on resources that were truly scarce, like the absorptive capacity of Earth's atmosphere, rather than focusing on reducing the need for low-skilled labour?[34]

None of this is to argue against technological progress. It after all sits behind human progress. But it is to argue that the incentives created by markets and government have a big impact on the *kind* of technological progress we get.

Getting your flavour of innovation

One of the most famous volumes on the economics of technological change is titled *The Rate and Direction of Inventive Activity*.[35] Yet the main focus of academics and policymakers has been on increasing the rate of innovation, how fast we innovate, while the direction, what it is we innovate on, is an afterthought. This is a mistake. The direction matters to the future of jobs, automation, climate change and inequality. Moreover, even when policy does not attempt to direct innovation, it ends up doing so anyway – just not always in helpful or efficient ways.

It is time to think much more clearly about the direction of technological change and how we might want to shape it. Some have already begun to do so. Christensen argues that we need to direct innovation to be more market-creating. Similarly, the late Sir Anthony Atkinson, the father of the study of inequality, includes influencing the direction of technological change as his very first proposal for addressing inequality, arguing that it 'should be an explicit concern of policy makers'.[36] And Acemoglu has led the academic literature on what he calls 'directed technical change', arguing that it can and should be shaped.[37]

The ability of governments to shape incentives and innovation is wider than often acknowledged. There are two broad options: first, shaping the incentives for the private sector to innovate and adopt particular kinds of technologies and, second, directly investing in research and innovation in particular areas. Let's take them in turn.

Profits and relative prices of inputs matter; the kind of innovation we get is strongly influenced by them both. Yet they are not created in a vacuum. Instead they are heavily influenced by policy choices about taxes, labour regulation, and missing or inadequate markets for things like carbon. Indeed, just about every investment, policy and decision that governments make in research, infrastructure, customs, education, environmental regulation, roads, trade deals, broadband and immigration shapes input costs and paths to profitability for firms in

New Zealand. This in turn influences the decisions of businesses and innovators about technological development.

Sometimes these are deliberate choices. As Phil O'Reilly, former chief executive of BusinessNZ, put it to me, 'If you don't think we pick winners, try taking an apple through customs ... We picked agriculture.'[38] Yet governments also shape markets in part because they must. Norms and regulations are required just to enable trade, facilitate insurance and protect consumers, to pick three examples from hundreds. Specific inputs are needed for each sector in a modern economy, some of which the private sector doesn't provide enough of, such as knowledge, research, and educated, healthy workers. There is just no escaping the fact that governments have to make choices and investments to help the modern economy function, and that these shape the direction of innovation. As Dani Rodrik and Ricardo Hausmann of Harvard University have memorably written, governments are doomed to choose.[39] If they are doomed to choose, they should at least choose smartly with close attention to how their choices affect innovation.

Reasonable people may disagree on how innovation should be directed: some may favour hastening automation in search of reduced working hours, while others may worry that moving rapidly could be unnecessarily painful. Yet one sensible step would be to regulate in ways that address missing markets for things like carbon, to help businesses make investment decisions that reflect what we actually don't have enough of. For example, one of Stiglitz's top policy recommendations for the future of work is 'a high carbon tax – to encourage resource saving innovation at the expense of labour saving innovation'.[40] A carbon price that reflects the limits we face would force everyone to pay the true cost of pollution rather than leaving the poor of Vanuatu and Bangladesh to bear the consequences. This would change the technologies that businesses would adopt and, in turn, which technologies innovators would try to create.

If airlines like Qantas and Air New Zealand faced the true cost of the carbon they emit, they would be in a heck of a hurry to find an innovation that could reduce their emissions. Researchers and innovators, along with the investors that back them, would know there was a lot more money to be made in innovations that reduce greenhouse gas emissions from air transportation. In such a system less attention would be given to innovations to reduce low-skilled labour like check-in attendants, and more time and money invested in innovations to reduce carbon emissions. In the long run, the world might have more jobs and fewer floods.

The evidence backs up the theory; innovation focuses on the most costly inputs. A study by Adam Jaffe, now the director of Motu Economic and Public Policy Research in Wellington, found that when energy prices were stable, innovations in air conditioners focused on reducing their price. However, when energy prices increased, air conditioners became more energy-efficient rather than simply cheaper.[41] High energy prices, whether driven by policy or markets, have a strong positive effect on the number of successful patents for technologies helping improve energy supply and energy efficiency.[42] Building on these studies, Daron Acemoglu and his co-authors have modelled how even temporary carbon taxes and research subsidies 'would be sufficient to *redirect* technical change and avoid an environmental disaster'.[43] Policy really can influence the kind of innovation we get.

Patent policy is another area that can sharply influence both the speed of technological change and what gets invented. Patents help inventors (and their backers) benefit from their idea before others can copy it by giving them the sole right to use it for a fixed period. Historically, there is good evidence that patent protection helped encourage innovation.[44] Today, there is robust debate about whether patents give inventors protection for too long – or for ideas that are too broad – thereby slowing down innovation by stopping others from quickly building on their work.[45] Prizes for innovations in certain areas, discussed in Chapter Seven, may be more efficient ways of influencing what inventors work on, allowing others to build more quickly on breakthroughs. There are other policy options too. To change the direction of technological change governments can also do things like accelerate the patent review and approval process for areas we really care about. The USA did exactly this for clean technologies.[46]

Innovators' incentives are also strongly influenced by other big policy choices such as tax policy, labour and monetary policy. These affect the relative price of labour versus capital. If they raise the price of labour relative to capital, as they often do, they make it even more worthwhile to invest in creating machines to replace human workers.

Labour in New Zealand, as in many countries, is taxed more heavily than capital.[47] Kate Raworth in her book *Doughnut Economics* points out that in 2012 over 50 per cent of tax revenue raised in the European Union came from taxing labour. In the USA it was higher still.[48] Meanwhile some capital investments are tax-deductible, there is no capital gains tax in New Zealand, and firms can finance machines with debt, which has further implicit subsidies.[49] Want to hire a human? There's a tax for that. Want to buy a robot? Here's a subsidy.

Support for capital investment may help generate valuable productivity improvements, but too stark a contrast between the taxation of capital and labour can mean businesses replace workers with machines even when it is not socially optimal.[50] And it can reward innovators and businesses for focusing purely on replacing workers with machines through efficiency innovation, rather than incentivising them to pursue market-creating innovations that require (and benefit) workers as well.[51] As Harvard economist Rodrik puts it, 'Too often, they [governments] subsidize labor-replacing, capital-intensive technologies, rather than pushing innovation in socially more beneficial directions, to augment rather than replace less skilled workers.'[52] In short, capital and labour tax policy can powerfully shape how much, how fast and whether businesses and innovators replace workers with machines.

How easy it is to switch jobs – labour-market flexibility – can also impact on the level of automation we get and therefore the incentive to invest in innovation to automate work. Acemoglu and Restrepo recently showed that the combination of relatively inflexible labour markets and subsidies of capital can actually result in automation that is *socially excessive*.[53] That is to say, these policies can result in automation that makes us collectively poorer than we would be if businesses had automated *less*. In such a case, automation would be making the national pie smaller. With the current hype around automation, that is a risk. Indeed, Elon Musk recently admitted that 'excessive automation at Tesla was a mistake'.[54]

What it costs to raise money to invest in machines can also strongly influence innovation. The last decade has also been one of historically low interest rates in developed countries. Since 2008, trillions of dollars have been injected into global markets by central bankers through quantitative easing. This has perhaps been the cheapest period in modern history to raise capital. Funds and investors worldwide have been desperately searching for good returns. It has never been easier to raise money to invest in a robot to replace a human. It has also made it easier for firms to raise capital for R&D if they want to.[55]

All these policies – tax, labour market and monetary – shape how profitable it is for businesses to buy a machine to replace a human and thus also how profitable it is to try to invent such machines. They directly shape the innovation that occurs.

There should be no illusion that innovation is responding to either a truly free market or a level playing field for workers and machines. You can argue about whether the policies that countries have on tax, labour markets and monetary policy among other areas are a good or bad idea based on their impact on innovation and automation. You could also advocate for policies that accelerate automation or slow it down.[56] If you wanted to accelerate automation, for example, you would want to see subsidies for capital investment, taxes on labour, and policies that keep returns high for innovations and that reduce the need for labour.

Yet, whatever your view, these policies do have a significant impact on innovation. Unfortunately, this basic truth is often ignored. Innovation and automation are instead depicted as a tsunami over which we have no control. Or, in a variation on the theme, it is suggested technological change is just the logical result of innovators responding to the signals of an unfettered free market. Neither is true. Both these ideas diminish our sense of our ability to shape our future – and therefore our likelihood of doing so.

Getting stuck into innovation directly

The state – the full weight of the government and public sector – is an influential thing. If governments want to push research and innovation in a particular direction they can, of course, pay for it themselves. States around the world do this all the time. Mariana Mazzucato, an Italian economist, has spent much of her career highlighting just how influential – and extraordinarily effective – states can be on innovation if they set their minds to it. Think your iPhone is all thanks to Steve Jobs' genius? Think again. In her book *The Entrepreneurial State*, Mazzucato shows through a dizzying array of examples that the state is in fact a real engine of innovation and has been behind many of the biggest technological breakthroughs in recent history, from the internet to wireless networks, to GPS, to touch screens, biotech and cleantech. GPS, for example, was developed almost entirely, and at major cost, by the United States Department of Defence, and is still maintained and improved by the US air force. Your iPhone is just one of many things that state-led innovation has helped make possible.[57]

The state can be publicly minded, wealthy and patient. All three make it an important innovator. State investment in R&D is crucial because knowledge is what economists call non-rivalrous; others can use it without stopping me from using it. For example, I can use my mother's recipe for muffins without preventing her from cooking a batch. The private sector left to its own devices tends to underinvest in new knowledge because it is difficult to capture the full benefit of the knowledge it creates – others can use it too. Patents address this, but imperfectly. Left to the private sector only, we would invest less in knowledge generation and research than would be best for society.

The state's sheer wealth matters too. In most cases, no single business has the resources to drive fundamental research and innovation at the same level as the state. In New Zealand, state-funded research via Crown research institutes and higher education is larger than the entire private sector spend on R&D.[58] In the USA, state research entities affiliated to the military have been major leaders of fundamental innovation. Given that state funding ultimately traces back to one place – the government – R&D spending can be focused on a particular problem or mission to powerful effect. That kind of scale and focus in research can be very difficult in the New Zealand private sector.

Perhaps most importantly, there is deep uncertainty and long time frames involved in innovation. Private companies can be unwilling to take on such uncertainty over long periods, and investors and bankers can be similarly reluctant to sign off on long, uncertain projects. As Christensen argues, generating market-creating innovations requires an intentional focus, a longer-term mindset and an appetite for uncertainty. The private sector tends to underinvest in this kind of long-term market-creating innovation, partly because of the obsession in boardrooms with efficient use of capital and short-term returns.[59] Efficiency innovation, which often lowers costs but eliminates jobs, is instead the focus.

States on the other hand can think and act over longer time frames and with a different attitude to risk. This enables them to both do more basic foundational research and give nascent innovations the time they need to develop into something genuinely new. Mazzucato suggests that the state should be understood as 'neither a "meddler" nor a simple "facilitator" of economic growth. It is a key partner for the private sector – and often a more daring one, willing to take the risks that business won't.'[60] Long-term innovation supported by the state has a good chance of being of the market-creating, and job-creating, kind. Many of Mazzucato's examples fit this well: the internet, GPS and wireless networks all had significant state involvement and were truly market-creating.

Christensen, ever the Harvard Business School professor, wants to see the private sector get back to pursuing longer-horizon market-creating innovation – and there are some changes that could help with that, which are discussed below. But for Mazzucato, that's a chimera; it's always been the state that is most able to play that role.

Of course, as Chapters Six and Seven will discuss, we may want a world of greater automation, and the state can also shape innovation in that direction using similar tools to prioritise efficiency innovation. However, if it's a future of high employment we want, then market-creating innovation may need to be a clearer goal for our innovation system.

The key to success in state-led market-creating breakthroughs, Mazzucato argues, has been 'mission-oriented' innovation policy. This is distinct from research and innovation for general advancement. It is 'big science deployed to meet big problems'.[61] Missions should be well defined, address a societal need or demand, and create a long-term public agenda for innovation policies. They are not just about picking sectors to support, or even stating the importance of a big challenge like climate change. They require tackling a specific problem with a specific target and time frame, and require multiple sectors to act. Mazzucato cites Germany's *Energiewende* (energy revolution) policy as a modern-day example of a mission with a clear target: to reduce greenhouse gas emissions by 40 per cent by 2020 and up to 95 per cent by 2050, compared to 1990 levels.[62] Reinforcing this, Germany has specific targets on the share of renewables, energy efficiency and phasing out nuclear power. Sectors across the economy, not just energy, have to reinvent themselves and invest in serious R&D to fulfil this mission.

Mazzucato shows that the state really can shape innovation. But she is also concerned that in many countries policymakers have forgotten their role. Have New Zealand's?

Innovating for employment in Aotearoa

Today one of the most important policy goals in New Zealand is continued high employment. This book will discuss a scenario in which that changes. But for now, as concerns grow about automation displacing workers, it is fair to ask whether our innovation efforts are in line with that goal.

We certainly talk the talk. We have a ministry focused on innovation: the Ministry of Business, Innovation and Employment (MBIE). Putting both innovation and employment in the ministry's title is a strong start. Such a focus on innovation is welcome, and we have made some progress. Yet in 2018 New Zealand was ranked 22nd in the world on the Global Innovation Index, down from 17th in 2013.[63] R&D spending is just 1.2 per cent of GDP.[64] Business investment in R&D is particularly low by international standards, in part because of the paucity of large firms in New Zealand, which typically invest more heavily in R&D. We also have a real diversity problem in science; despite women representing 50 per cent of graduates in science only 32 per cent of the scientific workforce is women. Even more alarmingly, just 2 per cent of New Zealand's science workforce is Māori or Pasifika.[65] There's a real risk that this narrows the range of challenges our science and innovation system tackles.

Successive governments have made raising business expenditure on R&D a priority, and a range of public initiatives encourage business spending on R&D or co-invest with businesses. But what kind of innovation are we getting for these public investments and incentives to the private sector? Or, put differently, do the innovation and employment bits of MBIE talk to each other?

The primary sectors, oil and gas, and mining have been major focuses of government science and innovation funding via the Primary Growth Partnership, the Energy and Minerals Research Fund and Crown research institutes.[66] The primary industries alone in 2014 received $444 million in R&D spending. Of this, over 40 per cent was government spending.[67] Together this made the primary industries the sector with the most government R&D funding.

Yet Christensen's view is that innovation in resource-intensive industries tends to be efficiency innovation rather than market-creating innovation.[68] Many of the Primary Growth Partnership's current programmes fit this characterisation – they allow an existing process to be done more efficiently. For example, the Pioneering to Precision programme allows fertiliser to be better applied to hill country, and Steepland Harvesting aims at more efficient and safer harvesting of trees on steep hill country. The benefits efficiency innovation generates are important; these programmes help improve safety and efficiency. But equally we should be clear that direct job creation is not typically one of them.

Other parts of New Zealand's publicly funded innovation system are less focused on the primary sectors with their associated tendency towards efficiency innovation. Callaghan Innovation, for example, supports high-tech businesses from a wide range of sectors and helps drive commercialisation of new ideas. It has backed everything from the now famous Rocket Lab, which is creating New Zealand's very own space industry, to Fabrum Solutions, a company which is commercialising and manufacturing portable cryocoolers (high-tech refrigerators that can go as low as -250°C) to store liquid oxygen that can be used in hazardous situations, mining emergencies or air ambulances.

Yet New Zealand's publicly supported innovation system, including Callaghan Innovation, does not appear to be particularly focused on the kinds of initiatives that create new markets and new jobs. Instead it is neutral towards the job implications of the innovation it funds. The job-creating potential of a project is simply not a criterion for most of our innovation spending choices. For example, Callaghan Innovation's R&D Project Grant application template does not ask businesses for employment growth or job-creation projections. Similarly, the criteria for selecting R&D projects do not include employment or job creation. Indeed, neither the word 'job' nor 'employment' even appear in the Ministerial Direction setting out the goals and criteria for R&D Project Grants by Callaghan Innovation.[69] Similarly, the employment and job-creation potential of a project is not specifically assessed as part of decisions on funding from New Zealand's largest contestable research and science fund, the Endeavour Fund.[70] This, at a minimum, is a failure to align innovation policy with the oft-stated goal of every New Zealand government in my lifetime: job creation. This goal could change, but until it does, and considering fears about automation and jobs, our current approach seems odd. If we want market-creating innovation, which is most likely to create jobs, we have to aim for it.[71]

Instead New Zealand's innovation system appears to be heavily invested in efficiency innovations, often in carbon-intensive sectors. If job creation remains our aim in the future, we should add a filter to much of our publicly financed innovation spending: does this project offer the chance of genuine market-creating innovation with its greater possibility of creating new jobs? That's not an easy question to determine in advance – and there is a real risk such efforts can be gamed – but today we do not even try to answer it.

A clearly defined nationwide mission could orient New Zealand's innovation system towards market-creating innovation. These innovations often require a longer-term perspective and greater risk appetite, which a mission could provide. One step in the right direction has been the establishment in 2013 of the National Science Challenges. These eleven challenges have funding attached directly but are also the main tool to align the research funded through the Endeavour Fund, Crown research institutes and other funds.[72] Yet the National Science Challenges are still essentially areas for research with broad objectives, rather than missions with specific targets that can truly galvanise innovation and research across sectors.

On 25 May 1961 John F. Kennedy launched the research mission *par excellence*: sending an American safely to the moon by the end of the decade. What is the New Zealand equivalent in ambition, timeline and accountability? Moreover, in the future, if it is jobs we want, do our publicly declared missions hold out the possibility of creating new jobs?

What's the purpose of business anyway?

The state is not the only entity that influences innovation. Most people in New Zealand, and in rich countries around the world, work in private companies. Yet today the willingness of private companies to take risks on technology development and adoption are heavily informed by a particular mantra: maximise shareholder value. Corporate jargon is hardly a topic to get the heart racing. Yet maximising shareholder value deeply underpins the path of innovation and automation. If a world of work remains our idea of the good society, we need to re-examine how our corporate mantras might make attaining that more difficult.

First, though, what does 'maximise shareholder value' really mean? Simply, it means get the biggest damn profit possible for the owners of the company. This idea is so ingrained as to seem mundane these days. Yet it only gained prominence in the 1970s, driven by the swing to free-market economics at the time. As late as 1981 the United States Business Roundtable declared, 'Corporations have a responsibility, first of all, to make available to the public quality goods and services at fair prices, thereby earning a profit that attracts investment to continue and enhance the enterprise, provide jobs, and build the economy.'[73] By 1997 they had changed their view: 'The principal objective of a business enterprise is to generate economic returns to its owners.... If the CEO and the directors are not focused on shareholder value, it may be less likely that the corporation will realize that value.'[74] That's a heck of a change, and narrowing, in the point, or purpose, of business.

These views were echoed in New Zealand, where Business Roundtable executive Roger Kerr wrote in 1996 that corporate social-responsibility activities 'are not typically "social responsibilities" but remain discretionary acts for which managers should be held responsible only to the shareholders employing them'.[75]

Maximising shareholder value is pretty good for shareholders. What the mantra obscures is what it means for the people who show up to work every day: employees. They become just another production cost to be minimised for greater short-term profits rather than a real stakeholder in the company.[76] Thinking of employees as a cost, rather than as human beings with potential creative ways to improve the business, can make it all too easy to replace them with a machine.

This relates to the second big issue with the 'maximise shareholder value' mantra: short-termism.[77] Throughout this transformation in business thinking, the standard measure of shareholder value has been quarterly earnings reports. Financial analysts wait with bated breath for the latest earnings reports, and trade shares aggressively on their results. A bad earnings report can put executives at risk of the chop. This short-term focus leaves businesses less likely to take a long view of the value of workers based on their creativity, institutional knowledge or ability to retrain and do new things. Instead, if they can be replaced by a cheaper machine, well, you have to maximise shareholder value for the next quarter's earnings report.[78]

Seeing workers primarily as a cost and focusing on the short term can have a flow-on effect on innovation. If every executive is looking for ways to cut costs to improve margins and hit their quarterly targets, *buying* a machine that can replace workers is a great use of time and money. Corporate decisions like these – even if not in the long-term interests of the business – echo quickly beyond individual companies. They spread along supply chains to teams of innovators, researchers and product developers around the world and, in doing so, mean that *inventing and developing* a machine that can replace workers is also a great use of time and money. In this way, the innovation system can become skewed towards efficiency innovations that replace workers.

This is exactly the concern of Christensen and van Bever; the corporate obsession with short-term results distorts the way capital is invested. In their view, it leads companies to invest in efficiency innovations more heavily. Worse, it points them away from market-creating innovation, which could result in a new product or way of doing business that could grow employment, because this kind of innovation tends to be riskier and slower.

The focus on maximising shareholder value by business managers has both shaped, and been shaped by, investor impatience. If patience is a virtue, virtuousness is in short supply. A study from the Bank of England finds investors' focus is excessively short-term compared to what would be most profitable. To be clear, this is not good business; it is reducing investors' returns in the long run. Moreover, investors are becoming even more impatient over time.[79] The median holding period for shares is about ten months.[80] With impatient investors, even business managers that want to make decisions for the long term are in a tough spot.

As maximising shareholder value became more dominant in business, attention turned to how to ensure managers actually followed the mantra. Economists had a clear answer: align the incentives of the shareholders with the incentives of business managers. In practice this meant paying top managers more and more of their income in shares of their company. This has had further worrying effects on innovation.

Executives around the world have been spending their company's cash in record amounts on repurchasing the shares of those companies. These share buy-backs are often massive in scale and boost the share price. The higher the share price, the more managers' own pay packets are worth. You can see why share buy-backs might be tempting.[81] In the USA that temptation seems to have been overpowering; share buy-backs on the S&P 500, an American stock market index, were set to reach US$900 billion in 2018. That's nearly double the previous few years, which themselves had been bumper years.[82] Share buy-backs in New Zealand are smaller, but they are still frequent – to the concern of some observers.[83]

The issue is that spending US$900 billion on share buy-backs means that US$900 billion is not spent on other things, like R&D for market-creating innovation.[84] Nor is that US$900 billion spent on paying workers better wages.[85] From an innovation and automation perspective this can leave workers with the worst possible outcome: the long-term investments that are needed for market-creating innovation, which helps create jobs, are avoided in favour of short-term investment in efficiency innovation, which tends to eliminate jobs, or simply in favour of share buy-backs.

Doing business like the future (really) mattered

This version of capitalism, however, isn't the only show in town. If a future of high employment is our goal there are alternatives that could support more market-creating innovation and job growth. (If, on the other hand, a world of far less work is our hope, then we should embrace further a strong-minded focus on efficiency among business. Chapter Six will tackle this approach directly, but in either case our choices can shape the kind of technology we get.)

Questioning the nature of modern capitalism can get you quickly tarred as economically illiterate or naïve. Yet Dominic Barton, until recently managing director of the global consulting firm McKinsey and Company, has heavily criticised the short-term focus of capitalism. His research found that short-termism in business is increasing, but that, actually, firms with longer time horizons do better.[86] Revenues of businesses with a longer-term focus were up by 47 per cent on other firms. They also had stronger financial performance. They spent 50 per cent more on R&D than firms that were more short-term focused, increasing the chances of market-creating innovation. And, notably, they added nearly 12,000 more jobs than other firms on average from 2001 to 2015. Quarterly earnings might not be what they are cracked up to be.

A first step toward an alternative, as Barton has argued, is to get rid of 'quarterly capitalism'. Start by chucking out the practice of quarterly earnings reports. Sound too radical? Coca-Cola, Ford, Unilever and Google have already done so.

Yet for this to become more widespread the big beasts of financial investment – pension funds, mutual funds, insurance companies and sovereign wealth funds – must focus on the longer term too. That means accepting more risk of short-term underperformance for long-term gain and judging their own performance against long-term benchmarks.

Berkshire Hathaway, Warren Buffett's famously successful investment company, does exactly this.[87] The New Zealand Superannuation Fund is another great example. It declared itself a long-term investor when it began in 2001 and its returns have been so good that the *Financial Times* dubbed it the All Blacks of funds.[88]

A longer-term perspective from institutional investors helps the companies they invest in to think longer-term too.[89] Still, money talks. If we really wanted managers to focus on the long term, boards could link executive pay to measures of long-term progress rather than tying pay to share price.

Broadening our idea of the purpose of business beyond just maximising shareholder value would be a powerful second step. Eric Beinhocker, director of the Institute for New Economic Thinking at Oxford, and American entrepreneur Nick Hanauer argue that 'providing them [shareholders] with a return that is competitive compared with the alternatives is a boundary condition for a successful business; it is not the *purpose* of a business. After all, having enough food is a boundary condition for life – but the purpose of life is more than just eating.'[90]

There are also lessons to be learnt much closer to home. The Māori values of whanaungatanga (kinship) and kaitiakitanga (guardianship) influence Māori approaches to business and embody the importance of thinking longer-term and about broader stakeholders. As Diane Ruwhiu of the University of Otago told me, at its best in Māori business there is an 'accountability that goes across generations; iwi making decisions today are accountable to their ancestors and to their grandchildren'.[91]

A broader approach drawing on whanaungatanga and kaitiakitanga could start by giving greater weight to the concerns of employees and customers. Business, after all, relies on public trust for its very existence. A 2018 study found that only 47 per cent of the general population of New Zealand trust business.[92] This is lower than the global average of 52 per cent and well behind places like Singapore, 56 per cent, and the Netherlands, 60 per cent.[93] Job cuts due to short-termism hurts this trust. In New Zealand, 43 per cent of us believe that ensuring there are plenty of available job opportunities is a mandate for business. Yet just 28 per cent believe business is achieving that mandate. A change in approach to focus more on customer and employee concerns would mean that business needs to move beyond efficiency innovations into developing or adopting market-creating innovations that could bring in new customers and create jobs.

Treating workers as more than just a cost can in fact help business perform better and innovate. 'Workers and the knowledge workers hold is too often treated as dispensable,' says Rana Foroohar, the global business columnist for the *Financial Times*. 'Nurturing human capital and then harvesting its riches in the form of intellectual property is really a hallmark of today's most successful businesses.'[94]

Foroohar is right, but the value of empowered workers isn't just true for high-tech firms like Microsoft, Google and Facebook. The same holds for regular old retailers. Zeynep Ton, an academic at MIT and head of the Good Jobs Institute, looked me sternly in the eye and told me that it's simple: treating workers well pays off. Her research shows that retailers that pay higher wages, train their workers and provide more convenient schedules can actually do better.[95]

This is not a niche experiment unproven at scale. Companies successfully following what Ton calls a 'good jobs strategy' include Costco, an American wholesaler with over 580 stores; Mercadona, Spain's largest supermarket chain; Trader Joe's, an American supermarket with around 340 stores; and QuikTrip, an American convenience-store chain with more than 540 outlets. These companies' workers get more done every hour and make better use of their stores and warehouses. Their employees also tend to stick around – avoiding the costs for companies of frequently hiring and training people. Strikingly, these companies do all this while competing successfully on *price*.

The key point, Ton emphasised to me, is that employees do not have to be short-changed for business to deliver low prices. In fact better-paid, more empowered employees who are trained to do a wide range of tasks spend their days quite differently to other retail employees.

They solve problems, reduce stock wastage, figure out ways to improve displays, charm and help customers, and use their initiative to find savings whilst increasing sales. Naturally, it makes these jobs a lot more meaningful too. Poor treatment of workers may not just be bad for workers; it is leaving money on the table. With such a change in attitude to workers, business would be wary of unthinking automation, which can cause job losses, and innovators would focus less on inventing new ways to replace workers.

As Ton told me, if you can succeed with a good jobs strategy in industries that compete on low prices, you can succeed with it in industries that are focused on customer experience and quality. Listening to her, it was hard not to think of the many New Zealanders in jobs in supermarkets, hospitality and horticulture (to name just a few) that don't pay enough, nor challenge or inspire them.[96] A good jobs strategy in many New Zealand companies could improve lives right now, remind us all of the value of empowered workers, and give us pause before engaging in excessive, cost-cutting automation.

If there are such advantages to a good jobs strategy, why do so many companies still favour paying workers less and replacing them with machines to improve short-term profits? A good jobs strategy can get companies to a place where they are more profitable and their workers are better paid and more empowered. But getting such a strategy right isn't easy. It takes an understanding of the evidence, time, effort, training and a mindset shift across the company. That can be a tall order in a world of quarterly earnings reports. Consequently, many companies settle for doing worse and treating their workers as a cost to be minimised or cut.

Corporate governance: two words so dull your publisher warns that you are about to lose half your readership. As surely every business-school student has said, who cares? But maybe that's the point. A topic so dull it puts you to sleep. Yet corporate governance, in all its bland beige boringness, decides who has the power. So while you doze, the decision to replace workers with machines has been taken. Want to have the whip hand, captain the ship, hold the aces? Corporate governance determines how and who gets the whip, captain's hat and aces.

Germany and Japan have been strikingly successful at generating high-wage jobs, in manufacturing especially. Both give workers an active role in corporate governance; German corporations have worker representation on corporate boards. Japanese firms regularly consult with workers as part of *kaizen* – the ethos of continual improvement that has underpinned much of Japan's corporate success, notably Toyota's.

Changes to corporate governance can also directly target short-termism. Shareholders can exercise power through voting on strategy, pay and plans at annual meetings. Yet someone who has held a share for a hot minute and will sell it tomorrow has the same vote as someone who holds for the long term.[97] If you gave voting rights to every tourist in New Zealand at election time they might not vote in New Zealand's long-term interests. The same is true for companies. Yet this could change; some companies in France give two votes to shareholders who have held shares for more than a year.[98]

The extent to which employees are on boards, the time frames and goals businesses have, and their attitudes to workers all influence how much businesses invest in automation to replace workers, compared to what they invest in new innovations that could instead create jobs. These investment priorities in turn influence what technologies the innovation system tries to create. That's worth making sure we get right.

For now, we shouldn't lose sight of the fact that not all innovation is the same. All of it can enhance productivity, but efficiency innovation is likely to reduce the number of workers needed to produce a product, sustaining innovation helps maintain jobs, and market-creating innovation can directly create new jobs.[99] The type of innovation we get at scale, and how much of it, is not fixed. Instead it is determined by the incentives innovators face and the support they get.

These incentives are directly shaped by what governments do, how our businesses choose to operate and how we as consumers and citizens respond. The relative prices of different inputs to production matter a great deal: innovators try to find ways to replace the pricey ones. In rich countries, that is often workers' wages. Yet that's no function of perfect markets. It could be much more worthwhile economising on carbon, charged at its true cost. Tax policies around the world favour capital, making it more profitable to invest in machines rather than humans. This in turn increases profits for the inventors of those machines and encourages more of the same. Super-low interest rates by Reserve Banks grease the wheels further. Moreover, the way we set up our companies often leaves workers viewed as a production cost to be minimised, again making it profitable to invent machines to replace them. Depending on the future we want to head for, this can be helpful, or very bad news indeed.

The point is that none of this *has* to be this way. Depending on the future we want to create, there are alternatives. There are other ways to run companies. Monetary policy can focus more on reducing unemployment. Tax policy is ultimately decided by a democratic process and need not favour automation. Governments could push their innovation systems toward market-creating innovation far more than they do; as Mazzucato shows, they can be very effective at doing so when they set their minds to it. As for any policy choice, there are real trade-offs, some larger than others. But there is no doubt we have far more influence, and potential control, over the path technology takes than we are often told.

Two
The proof is in the use

The economic power of innovation comes mainly
from its adoption, not from its creation.
– Fredrik Erixon and Björn Weigel, *The Innovation Illusion*

In 1883 Charles Fritts, a stylishly bearded New York inventor, created the first solar panel. It was magic: sunlight and then, tah-dah, electricity. Just a year later the first solar array was installed on a New York roof. Fritts predicted that 'we may ere long see the photoelectric plate competing, in high efficiency of conversion, with the dynamo electric machine itself'.[1] The first coal power plants had been built only three years earlier by Thomas Edison. In 1954 a team at Bell Labs in the USA created a solar panel using silicon rather than selenium – the same approach used in today's solar panels. The Bell team's announcement caused quite a stir. The *U.S. News & World Report* enthusiastically wrote: 'Fuel Unlimited: The [silicon] strips may provide more power than all the world's coal, oil and uranium ...' The *New York Times* piled on too, writing on page one that this invention 'may mark the beginning of a new era, leading eventually to the realization of one of mankind's most cherished dreams – the harnessing of the almost limitless energy of the sun for the uses of civilization'.[2] The future was bright, and lit by solar.

Fast forward to 2016 (a whopping 133 years after Fritts and a still chasmic 62 years after Bell Labs) and solar energy represented just 1.3 per cent of global electricity generation.[3] The Bell team, as well as the excited journalists, surely would have expected more in well over half a century. Fritts would be twisting his beard in horror at how slow we have been. The basic technology after all was on a New York roof in 1884.

Yet Rob Nail, chief executive of Singularity University, derides the International Energy Agency (IEA) for consistently underestimating the uptake of solar in their projections. Solar installations are way up, he points out in his talks. And the IEA keeps underestimating solar's growth, he notes, showing a graph of the IEA's conservative projections of solar growth since the early 2000s and the much more rapid actual growth. He is right. Growth in solar installations has been impressive since the mid-2000s. And the IEA does keep underestimating. For Nail, the lesson is crystal clear: we are underestimating the speed at which today's new technologies will become commonly used, just as we underestimated solar.

The only problem with this thesis, of course, is that for about a hundred years we did not underestimate solar. Instead, Fritts tossed and turned in his grave in frustration. Telling the story of solar as if it began in the early 2000s, over a hundred years after the first solar panel was installed, is cherry-picking the data. Much like glory-hunting sports fans that begin to support a team only after it has been bought by billionaires, this skips the years of struggle and goes straight to the triumph, acting as if success is easy and history only began yesterday.[4] For solar energy, the equivalent of a billionaire owner also showed up: governments worldwide provided billions in subsidies via very generous feed-in tariffs for solar electricity. This selective history of solar energy allows for the claim that we underestimate just how quickly new technologies can hit the big time.

Yet the story of solar, rather than suggesting that we are all underestimating new technologies' likely speeds of impact, reveals something else: there is an enormous difference between technically doable, commercially viable and, finally, commonly and productively in use. Solar power didn't fail to arrive more quickly or widely because it was a bad idea. It is a great idea that we desperately need. Instead, challenges of development, commercialisation and the entrenched interests of the energy industry played a big role – as did, in its recent spike in uptake, massive government support. Being technically doable and a great idea alone are not enough.[5] Despite this there is still a tendency to assume new technologies are the equivalent of early-2000s solar, rather than Fritts's 1883 solar or Bell Labs' 1954 solar.

It has become popular recently to quote the late American futurist Roy Amara's line that 'we tend to overestimate the effect of a technology in the short run and underestimate the effect in the long run'.[6] That's wise advice. But how long are the short run and the long run? In the case of solar we overestimated the effect for 120 years and now we have finally started to underestimate it. That's a pretty long short run.

Even today's dramatic technologies that appear to be arriving suddenly have often been under development a long time. Self-driving vehicles were already whizzing along at 90 kilometres an hour in 1987 and by 1994, twenty-five years ago, were driving around Paris on a three-lane highway in heavy traffic.[7]

There's also another possibility that is rarely considered. Some new technologies may simply be the equivalent of Apple's Newton (a digital handwriting message pad) or Trump: The Game (Trump-branded Monopoly). Both of these now grace Sweden's Museum of Failure.

This chapter will argue three big things. First, that adoption and diffusion are absolutely fundamental to how future technologies change work. Second, that in New Zealand technology adoption is relatively slow – but that it could be faster (or slower) with different policy and business choices. Third, that to think clearly about how technological adoption will happen in the future we need to keep in mind some core rules of thumb. Let's start with the question of adoption.

Half of all US jobs are at risk of automation. That statement, a pretty scary claim, was reported across the world in 2013. That was the year Carl Benedikt Frey and Michael Osborne of Oxford University published a study on the susceptibility of jobs to computerisation that claimed that 47 per cent of US jobs were at high risk of being automated within the next ten to twenty years.[8] It spawned thousands of ominous headlines, which alarmed not just me, but people around the world. The study has been cited 3,128 times according to Google Scholar, and touched off a bout of collective anxiety about the future of work that shows few signs of abating. Similar studies abound. In 2015 the New Zealand Institute of Economic Research, using the same methodology, found that 46 per cent of New Zealand jobs are at high risk of computerisation by 2035.[9]

Yet once you start to think seriously about the process of technology adoption, Frey and Osborne's study, and others like it, look a lot less scary.[10] Chapters Three and Four will detail the full array of problems with studies like this and suggest how we can think more clearly about the challenge ahead (and there is one). But let us look right now just at how these studies treat the adoption of technologies and thus understand what they are *really* saying.

If your boss said there was a high risk of you being fired you would be extremely worried. High risk sounds very likely; definitely more likely to happen than not. But what does it mean in these studies for a job to be at high risk? Your job is 'at high risk' if, according to the experts Frey and Osborne surveyed, it's likely that state-of-the-art computers could complete the tasks you normally do.[11] The implicit, and critical, assumption in how these results are often presented is that if a job *could* be automated, it *will* be. That is a shockingly big assumption.[12] The gap between technically automatable and actually automated at scale is large enough to send a fleet of self-driving trucks roaring through.

Technically doable is only the first step. Fritts's 1883 solar panel would almost certainly have qualified. Bell Labs' 1954 panels definitely would have. Technically feasible usually means someone somewhere, perhaps in a well-resourced laboratory, was able to do it. A bit like Fritts in New York in 1883. In the Frey and Osborne study, computerisable does not necessarily mean any machine has ever actually done it, just that experts believe computers *could* do it.[13] It does not mean that machines can do it at a cost that is competitive with alternatives; nor that people will actually want to use the technology, even if it is cheap enough. Nor that they will find a way to integrate the computer or machine into their daily business or life, nor that regulators will allow or facilitate machines doing that task. Finally, nor does it mean that the new technology can overcome the entrenched interests of the owners of existing technologies who want to protect the value of their assets, such as oil rigs, coal mines and gas power stations in the case of solar.

Ignoring the question of cost is perhaps the worst omission. As a result, when these studies say jobs are 'at risk' what they really mean is that they are at risk *if cost were no object*. Here are some other things that would be at risk if cost were no object: homelessness in New Zealand, child poverty, world hunger. We technically know how to build houses, feed kids and send food aid. But cost matters. This may seem an extreme reaction, but when reports are splashed in the newspapers saying that millions of jobs are at high risk solely on the basis of technical feasibility, without considering commercial viability (or anything else), they are doing the intellectual equivalent of holding the front page to say we could end homelessness in New Zealand if the state gave everyone a free house. Misrepresenting the nature of the risk from automation could have very real consequences; political decisions on spending, education, research and welfare can go in funny directions under the influence of headline-grabbing studies.

The Frey and Osborne study is a particularly egregious example of this way of reporting results, but essentially ignoring the question of technological adoption and cost-competitiveness is a common pattern in studies.[14] Then the media grabs the headline 'number of jobs at risk', and ignores any caveats about the speed of adoption and diffusion. Neither the studies nor the media coverage do the public any favours.

Why we should take a deep breath about robots replacing us

Am I over-egging the pudding by claiming that details like cost and technology diffusion matter greatly to the future of work? I asked MIT economics professor Daron Acemoglu. His answer, in short: no. He told me there are 'technologies that are just very difficult to turn into productive use', adding that 'large-scale automation is a bit like that. You know, it is very difficult to do it in a cost-effective way without making your production line more cumbersome.' He noted that in particular it 'may well be very difficult for medium-sized firms to solve this problem'.[15] For New Zealand the last point really matters: small and medium-sized enterprises make up 97 per cent of all enterprises, 29 per cent of all employees and 28 per cent of GDP in New Zealand – much higher than most rich countries.[16]

Even incredibly clever new technologies don't make it big on their own. People must decide whether to invest in and use those technologies. And getting to the point where new foreign technologies make a major difference – good or bad – to Kiwi jobs takes many steps. First, someone needs to bring technologies developed overseas to New Zealand, having decided they could be useful back home. Then other businesses must copy and learn so that business by business, person by person, new technologies spread throughout New Zealand from Auckland to Ōamaru and beyond. But having a new tool isn't enough. As Acemoglu points out, businesses and people need to find ways to use these technologies to enable them to produce more with less – to be more productive. This can be surprisingly challenging. It is only when technologies improve productivity that workers are affected in ways that can be positive, such as improved wages and job satisfaction, or negative, such as job losses.

The evidence of slow and uneven technological adoption and ineffective usage is easy to spot in New Zealand: the taxi driver still navigating using old TomToms rather than Google Maps or Waze, which provide real-time traffic information; most Kiwi dairy farmers getting up at 4.30 a.m. to milk cows when fully automated milking machines have been commercially available since the early 1990s; the fact that no one, except dorky tourists, uses Segways; the drone on the farm used more for fun by the kids than for checking on fallen fences; the disastrous adoption of the Novopay software system for paying teachers (which cost $45 million to fix); the continued use of a bus and driver to transport people from the domestic to international terminal in Auckland whereas at London Stansted and Frankfurt airports a driverless train does the job; the fact that Ports of Auckland is automating its container terminal in 2018 while Rotterdam did it in 1993 (and that most American ports have not done it at all).[17] Views may differ over whether quicker and more effective adoption of some of these technologies would be good or bad. But either way, technology diffusion is not a trifling matter.

Still, you might say, casual observations can mislead: show me the data. We know that countries that use more computers and machines, and in particular those that use better quality computers and machines, are more productive. Productivity, how much an average worker produces per hour, is thus a broad measure of how far new technologies have spread and how effectively technologies are being used. It is also a guide to what's happening to workers – when a new technology shows up in productivity statistics it is already changing work in a variety of ways by enabling businesses to do more with less.

Productivity varies enormously across countries. Even within the rich-country club of the OECD the average Irish worker produces over four times more per hour than the average Mexican worker.[18] Differences between OECD countries and poorer countries are far more extreme. Even in countries that appear very similar, the differences are stark. Each hour, the average worker in New Zealand only produces about two thirds of what the average worker in Australia produces – despite the strong similarities between the two countries.[19] Rugby is possibly the only exception where our productivity over eighty minutes tops theirs. Even in the exact same industries, the Australians produce more per hour than we do. The list of industries where this is the case is long: mining, agriculture, construction, retail, financial and insurance services, and most branches of manufacturing.[20] A big part of this difference is because businesses in New Zealand use fewer machines and computers than Australian businesses do, and the machines we use are not as good (economists call this capital intensity). Second, even when we use machines we don't manage to produce as much as the Australians do (economists call this multi-factor productivity because it controls for both workers' effort and the machines they have). This is not because we are less educated, but because we don't use our machines as effectively as they do due to things like our management practices, branding and economies of scale.[21]

The fact that countries like Australia are a lot more productive than New Zealand but still have high employment tells us something very important; while it is very hard to end up with widespread technological unemployment without rapid productivity growth, it is possible to have higher productivity without ending up with widespread unemployment. This is because the introduction of new technologies often helps to create new jobs, as Chapter Four will discuss in detail. Productivity growth also makes us richer and tends to improve wages. Yet as we saw in the last chapter, if we see a lot of efficiency innovations and few market-creating innovations, productivity growth *could* lead to higher unemployment. Much depends on how the economy and policymakers respond, as later chapters will discuss.

For now, though, we should start with the facts. And the facts are that New Zealand is just not very good at adopting technology and productivity improvements in general. That should give us all pause amid the idea that robots are soon about to sweep in for our jobs. We don't learn very quickly from what the best businesses are doing overseas. Even the crème de la crème of businesses in New Zealand produce about a third less per hour than the best businesses globally in the same industry.[22] We are ineffective international copycats. Not only that, we are getting less effective at copying each year. The best businesses globally are taking up new technologies and improving their productivity faster than the best firms in New Zealand. In productivity terms, today it is 'mind the gap' but in the future it could be 'careful of the chasm'.

Technology adoption is particularly slow in New Zealand, but others struggle too. An OECD flagship report, *The Future of Productivity*, has an ominous headline conclusion: the diffusion machine is broken.[23]

They find that workers at the world's most productive businesses are on average ten times more productive than at ordinary OECD businesses. This is because they use more and better machines and computers, they have more skilled workers and they have better ways (through management, marketing, network effects and other things) to produce more with the workers and machines they have.

Once a business brings a new technology to New Zealand, do other businesses notice and do the same? Not exactly. There are big gaps between the best companies in New Zealand and other Kiwi companies in the same sector.[24] Manufacturers are a little better at learning from each other than service providers, but overall we are ineffective domestic copycats too. In most countries less productive businesses represent a small share of the market and lose out over time. This makes sense: they need more people and more money to produce *less* stuff. Yet in New Zealand relatively unproductive firms remain big employers that also use a lot of capital. If putting new technologies rapidly into productive use were a running race, New Zealand would not be Peter Snell; we would be huffing and puffing back with the stragglers.

The bottom line is that while we worry about newly invented robots sweeping in and taking jobs, we in New Zealand do not actually use that many of the machines, computers and robots that already exist and are used elsewhere. Technological change will continue, but if history is anything to go by, its arrival in New Zealand workplaces may be relatively slow and halting.

What's so difficult about using new technologies anyway?

At face value, the fact technologies are often slow to spread is a puzzle, the reality contradicts what we might expect in theory. If a new technology improves how much a business can produce each day, businesses should rapidly invest in it so as not to lose out to their competitors. And if they do not get hold of new technologies, they should struggle. How can a business survive for long if it pays more wages and spends more money to produce the same amount of stuff as a rival company? Similarly, economic theory suggests that businesses will replace a worker with a newly invented machine if this works out cheaper in the medium term – but this is not always borne out in practice.

Wages vary dramatically between countries. By one measure, the average hourly wage in manufacturing in Switzerland in 2016 was US$60.36, whilst in the Philippines it was US$2.06.[25] Loans, on the other hand, are far more readily available and cheaper in Zürich than in Manila. Choices about whether to use technology are influenced by these costs. It doesn't make as much sense to invest in heavily automated production in Manila when wages are so low and it is expensive to borrow to invest. Yet, while important, this doesn't explain as much of why New Zealand might have far lower productivity and less use of cutting-edge technology than other OECD countries like the USA and Australia, given these countries have more similar wages and costs of capital.[26] What else is going on?

Ricardo Hausmann, the director of Harvard's Center for International Development, argues that technology does not diffuse quickly because of the nature of technology itself. As he put it to me:

> Technology is just an answer to a question of how to do something. And how do you do it, well, you do it by using tools, you do it by following procedures, processes, protocols, recipes and so on. But in order to implement technology, you have to mix those tools and those procedures with know-how, and know-how is something that exists only in brains.[27]

Know-how is tacit knowledge and is indeed very hard to transfer from one person to another.

Dan Carter's ability to ping over a drop goal from more than 40 metres out as a 105-kilogram flanker runs at him is a form of know-how. He can do it and others cannot because he has the know-how. Like most top sports players, he can't really explain to someone else or write down exactly how he does it (there is a reason why post-match sports interviews are often so dull). I could read endless written instructions of how to nail a drop goal under pressure from 40 metres, but if I go out and try I will almost certainly miss. Learning to use a new technology productively can be similarly challenging. It requires know-how, not just the machine and the instruction manual. We implicitly recognise the importance of know-how all the time. As Hausmann likes to ask, what do you do when your tooth hurts? Do you: a) search the internet for websites on endodontic surgery, buy some tools and fix it yourself, or b) look for a dentist?

Without know-how, new technologies can be very difficult to use effectively. Lack of know-how therefore limits the speed and extent of technological diffusion. Imagine you want to play Mozart's string quartets. If you have two violins, a viola and a cello, you have the physical technology required. But if you don't have anyone with the know-how to play the instruments, your attempts will just upset the neighbours. If you have one violinist already, you might be able to train another. But if you don't have any violinists to start with, you are in big trouble.[28]

Yet producing Mozart's string quartets requires less know-how and a smaller team than many modern goods and services. For example, to design and manufacture the ATR 72 plane that Air New Zealand flies between my home town of Nelson and Auckland requires huge teams of people with deep and specific know-how. We simply don't have that know-how in New Zealand. This gets to the heart of the challenge. As Hausmann put it:

> Modern technologies function not just with one brain but with an enormous network of brains that know different things and that collaborate on production. So the challenge of diffusing technology is really the challenge of diffusing the know-how not of just an individual but of teams.[29]

New Zealand has succeeded in this challenge in some sectors, most notably in our world-class film industry. For most products we purchase, from cars to iPhones, we have no idea who, and how many people, helped create them. But movie credits provide a rare window into modern production and underscore Hausmann's point. Consider *The Lord of the Rings: The Return of the King*, one of New Zealand's most successful films. Incredibly, the credits last a full nine minutes and thirty-three seconds – the longest credits ever at the time – and include such obscure roles as wrangler manager and compositing inferno artist. After watching nine-and-a-half minutes of highly specific roles and names rolling past I can assure you: modern production requires huge, complex teams, with deep and diverse know-how.

Despite the leading roles for Peter Jackson, Fran Walsh and other Kiwis, a wide variety of actors, producers, artists and designers listed in the credits are not New Zealanders. Instead they were brought to New Zealand from abroad to work on the film. Even Enid Dalkoff, that compositing inferno artist, is an American, a UCLA graduate. To succeed in modern production, you need to bring in the technology and the know-how of people who have already mastered the technology. It was this combination of a wide variety of know-how in a large team – with crucial pieces brought in from overseas – that enabled *The Lord of the Rings* to be produced in New Zealand.

So know-how is key to technology diffusion and it exists in brains. But why doesn't technology diffuse more quickly? Quite simply, brains don't move enough, and moving know-how from brain to brain is a slow, intensive and stubbornly in-person process. This is why all the successful car producers from Detroit came out of Olds Motor Works, founded by Ransom E. Olds in 1897. Olds Motor Works subcontracted parts production to others. Through intensively working with Olds Motor Works, these subcontractors learnt how to make whole cars and eventually started their own companies. They included the Briscoe Brothers (Buick), Henry Leland (Cadillac and Lincoln) and the Dodge Brothers (Dodge). A key player in General Motors, William Durant, was a former Olds Motor Works employee and also founded Chevrolet.[30]

There is a similar story in Silicon Valley. Fairchild Semiconductor Corporation spawned hundreds of businesses that 'established Silicon

Valley as a world center of entrepreneurial activity and technological leadership ... its surviving combined progeny have been estimated to be worth over $2 trillion [USD].'[31] The know-how that staff of these firms possessed enabled them to use and innovate with new technologies far more successfully than anyone else.

To bring the latest technologies, bring the brains

New Zealand businesses and policymakers are not powerless in the face of all this. They can influence the speed of technology adoption or the broad types of technologies adopted. Adopting new technologies underpins productivity growth, which is fundamentally what makes us wealthier as a society. Most commentators, from the Productivity Commission to the OECD, are therefore trying to get the government to shift policies to help technologies and ideas arrive more quickly in New Zealand. The losses to New Zealand if we don't succeed in this can be significant. For example, we already have to work longer hours than Australians to earn the same amount. If we don't succeed in increasing the speed of technology adoption, this may become even worse, we may have to do more physical and manual work than Australians, and we will likely be poorer than they are – at least on average. Yet new technologies can also be better for some people than others. Heavy adoption of efficiency innovations with little effort to use the resources that frees up on market-creating innovations could lead to rising unemployment. Although, as Chapter Three will show, we are not there yet.

The value of different kinds of new technologies depends on the kind of society we are aiming for: the subject of the second half of this book. If jobs remain a priority in the future, we need to pay more attention to ensuring that incentives aren't skewed only to adopting efficiency innovations. For now, it is important to remember that much can be done to affect the speed and direction of technological adoption – either speeding it up or slowing it down.

Silicon Valley became a hotbed of innovation in part because Fairchild Semiconductor Corporation was located there and the know-how of its workers helped generate many other businesses in the same location. But New Zealand isn't located next to Silicon Valley or any other hotbed of global innovation, so what can we do? Well, as Hausmann told me, 'immigration, we find, is critical to technology diffusion'.[32] Diego Comin, a leading expert on technology diffusion at Dartmouth University, went one further. Immigration, he said, was 'absolutely critical'.[33] And most critical is highly skilled immigration from technologically advanced places.[34] The Huguenots, chased from France to Prussia in the 1700s for their Protestant beliefs, introduced new technologies to Prussia, such as cotton printing and silk spinning. A hundred years later textile towns in Prussia that had received an influx of the new arrivals were still much more productive than their neighbours.[35] Similarly, Jewish scientists fleeing Nazi Germany revolutionised American science. Patenting increased by 31 per cent in fields with Jewish émigrés.[36]

This kind of impact still happens today. A recent study looked at the effect of science, technology, engineering and mathematics (STEM) immigrants to the USA. Remarkably, the authors find that these immigrants could explain as much as 30 to 50 per cent of US productivity growth between 1990 and 2010. They also found that these foreign-born, highly educated immigrants cause wage improvements for both the highly educated and the less highly educated native-born population.[37] Immigration is the first big tool to influence technological diffusion.

In New Zealand, immigration's role in spreading technology has received little attention. Still, research by Wellington think tank Motu suggests that 'firms that hire more recent immigrants and more recent returnee New Zealanders tend to innovate more than other firms'.[38] The research indicates that it is the skills of these immigrants that matters rather than the fact that they are immigrants per se. If the impact of immigration on innovation and productivity is less pronounced in New Zealand, as Motu's research implies, our immigration policy may not be especially well targeted. We operate a points system but we do not successfully prioritise immigrants from technologically advanced places, those with strong experience in, say, green technologies, nor immigrants with a strong science and technology background.[39] For example, in 2015 to 2016 the top three occupations for skilled migrant residence visas were: chefs (860), general retail managers (675) and café or restaurant managers (598).[40]

If we want to accelerate the adoption of technologies or to shape which technologies get the best chance at widespread use in New Zealand, we need to think again about our immigration policy. The Edmund Hillary Fellowship, supported by the Global Impact Visa, is one effort to do this. It gives entrepreneurs, investors and start-up teams the chance to come to New Zealand for three years and support to build innovative, high-impact organisations and businesses. Among the first two cohorts are entrepreneurs that have come to New Zealand to work on new forms of digital medicine, to use artificial intelligence to reduce waste in manufacturing, and to work on clean-energy grid integration.[41] That's a step towards faster arrival of new technologies, but in the longer run, special visa schemes like this could also assess applicants on whether their proposals might create markets and jobs. We may not think of it as such, but Immigration New Zealand is one of the key government agencies that will shape how technologies wind up affecting New Zealand.

Foreign immigrants aren't the only source of knowledge about new technologies overseas; New Zealand also boasts a diaspora of over a million. This isn't just long-disconnected Kiwis overseas passing their citizenship on to their foreign-born children who otherwise support the Wallabies or follow baseball. Fully 14 per cent of Kiwis who were born in New Zealand now live abroad. Among rich countries that is second only to Ireland.[42] Much of this diaspora lives in cities (and works in businesses) that are highly productive and technologically advanced, like London, San Francisco and New York. Members of this diaspora, if and when they return, bring new ideas, technologies and approaches to New Zealand. AnnaLee Saxenian, the dean of the University of California Berkeley School of Information, has documented in her book *The New Argonauts: Regional Advantage in a Global Economy* that Chinese, Taiwanese and Indian workers in Silicon Valley who have returned to their home countries have taken their know-how, networks and money with them and now play hugely important roles in developing high-technology industries in their home countries.[43]

I was extraordinarily fortunate to spend my early twenties surrounded by the 'Silicon Fens' at the University of Cambridge, and my late twenties studying at Harvard. Both experiences were eye-opening for the sheer array of cutting-edge work, and minds, around me. (My failure to launch a unicorn start-up is attributable only to my own failings rather than any lack of opportunity to see cutting-edge work.) We should be proud of New Zealand universities, which succeed in attracting students from abroad. But much of the next wave of work- and business-altering technologies are being invented in and around the world's leading universities in Europe, the USA, and, increasingly, Asia. If we want to speed up the arrival of those technologies in New Zealand much depends on whether there are any Kiwis around to soak it up, engage with the latest research and technologies, and bring them and the necessary know-how back to New Zealand.

Yet the classic Kiwi overseas experience rarely includes study. Only around 2 per cent of New Zealand's tertiary students are studying abroad.[44] While this beats the Australians, we are well behind many technologically advanced countries such as Sweden, which has 4.0 per cent of their tertiary students abroad and South Korea with 3.3 per cent of theirs. For both of these, the USA is the top destination. Malaysia, which is rapidly trying to move from middle-income to rich country status, sends 5.1 per cent of its tertiary students abroad, with the United Kingdom the chief destination. In this case, distance really isn't an excuse for us Kiwis. These countries have much larger student populations abroad despite the huge distances to their prime destinations.[45]

Permanently moving people and their know-how accelerates technology diffusion, but there are other possibilities. Companies spend US$1.2 trillion a year – 1.5 per cent of world GDP – on international business travel. That spending is growing at 6.5 per cent per year, almost twice the rate of global economic growth.[46] But why does anyone travel for business any more, now that we have top-notch video conferencing and free FaceTime and Skype video calls? Somehow, it is still much easier to get know-how to move from brain to brain when you are face to face. In person, you can choose where you look, pick up on broader sentiment in the room, catch the pre-meeting small talk while the microphone is off, and get the inside story over a beer in the evening. Businesses aren't spending all that money on travel for no reason.

Businesses that are plugged in overseas, whether through exports or foreign owners, also help technologies to move. Businesses that export compete with, and learn from, international rivals, who are often using the latest technologies. Kiwi exporters can't afford to ignore their foreign rivals' advances and so they improve too. Meanwhile, foreign owners smooth the path for new ideas and technologies to hotfoot it directly into New Zealand. In New Zealand, foreign-owned firms are especially productive, and exporters are more productive than our other firms.[47] The productivity of these businesses is also growing more rapidly than other New Zealand firms. Yet only 8 per cent of New Zealand firms export, while 1 per cent of New Zealand firms are foreign-owned. Foreign ownership can have downsides too – we at least want to be sure their productivity isn't a result of sending profits offshore without paying taxes. Yet in New Zealand few overseas business connections means that the adoption of technologies and productivity improvements are less widespread than they might be.

Distance explains some of this story for New Zealand. But that just means that if we want technology to arrive in New Zealand, we have to do more. If we want to accelerate technological adoption but also support new job creation, New Zealand Trade and Enterprise could be funded to support more businesses than the 700 exporters that are currently their focus. Foreign direct investment (FDI) is more controversial. Still, a smoother process for FDI in some important sectors like green energy investments could help technology adoption.[48] When we do screen FDI, we could focus much more on the technology investors can bring and try to target FDI that is market-creating and job-creating.

To influence the speed of technology adoption in New Zealand there are plenty of levers to pull: sharpening our focus on high-skilled immigration, encouraging the circulation and return of our diaspora, scholarships for young New Zealanders to study abroad and come back, greater business travel, and greater support to exporters and FDI. Some of these levers can also influence the balance between different types of innovations – efficiency, sustaining and market-creating – that arrive and reach scale in New Zealand. However, even if we speed up adoption overall, the kinds of innovations actually put into use in New Zealand will depend significantly on New Zealand's broader policy environment, as the last chapter underlined. Taxes on carbon and labour, subsidies or taxes on capital, which areas of technology have strong state funding, regulatory attitudes to new technologies; over time all these influence what kinds of innovations are most profitable or relevant and thus are trialled and succeed in New Zealand.

Regulation in particular can deeply shape the type of innovations that get adopted. Domino's delivered the first-ever pizza by drone in New Zealand in part because 'New Zealand had the most forward thinking aviation regulations'.[49] Driverless cars are being tested in American states chosen in part on their regulatory environment. The extent of their use in New Zealand will depend significantly on what our politicians decide. Direct subsidies for new innovations can also make a world of difference; remember how important generous feed-in tariffs are for solar-panel installations. Or just ask electric-vehicle manufacturers. Substantial subsidies have driven the uptake of electric vehicles in Norway and Denmark to levels far beyond New Zealand's (we have around 4,500, Norway has around 150,000). Yet when Denmark reinstated registration taxes on electric vehicles, sales fell by 68 per cent the next year.[50] Similarly when Hong Kong got rid of subsidies, sales of previously popular Teslas, a brand of electric vehicle, fell to zero the next month.[51]

Learning to learn

Small is beautiful, wrote E.F. Schumacher in his highly influential 1970s book critiquing modern economics and globalisation.[52] Yet when it comes to diffusing technology within New Zealand, size matters. New Zealand's national market is small and many businesses, especially in services, operate at an even smaller scale. They serve the local Timaru, Manawatū or Gisborne market only. If a German company that sells across the European Union fails to take up a new useful technology, they will lose out to competitors who do adopt new technologies. That just isn't the case in Dannevirke. The pressure on the average business in New Zealand to take up technology to keep up with rivals is much weaker. Unsurprisingly, businesses that focus on local markets in New Zealand are less productive than firms with a national focus.[53] New Zealand also has one of the highest proportions of old, small firms of all rich countries. Facing greater competition can be painful, but small, unproductive businesses tie up workers and capital – of which New Zealand has a limited amount – in not-very-productive businesses. This can have big costs for the country as a whole.[54]

If we wanted to increase competition to hasten productivity improvements and technological adoption, there are still a few tools rattling around policymakers' toolboxes. They could, for example, try to make it easier for consumers to switch who they buy from – this might mean support for comparison websites, easier contract terminations, or specific things like allowing people to keep their bank account number when they move banks, just like we do with phone numbers. Improving infrastructure to better link small towns and cities would also mean businesses could more easily compete in nearby markets.[55] The national broadband initiatives have already been a step in this direction, helping businesses advertise and connect with opportunities further afield.

You've got to spend money to make money, said Plautus the Roman playwright. That's essentially what Reserve Bank Governor Adrian Orr recently told business: invest.[56] Yet business investment in New Zealand is below the OECD average, despite our population growing faster than the OECD average. That is part of the reason why the amount of capital (think machines, computers, tools) available to the average Kiwi worker is only half of what is available to a British worker.[57] One big culprit, according to the Productivity Commission, is high interest rates. Lower them, they suggest, and businesses would invest more in equipment and technologies. Another is that, thanks in part to our dysfunctional housing market and tax system, savings in New Zealand are invested far more into houses rather than into business equipment and machinery that would improve our productivity. This skew towards housing hit an all-time high in 2017.[58]

There are policy tools to affect the cost of capital for businesses, such as incentivising greater national savings, adjusting the official cash rate and making investment cheaper via the tax system. However, as we saw in the last chapter, making capital cheap relative to labour could tilt the playing field in favour of automation at the expense of workers so far that it becomes economically inefficient. Socially excessive automation like this would actually make us poorer than we could have been with less automation. We are still some way from that in New Zealand. Yet if it is a future with continued growth and jobs that we want, the challenge for policymakers and business is not just to make capital cheaper to invest in general. Critically, it is also to increase long-term investments, using some of the approaches outlined in the last chapter, to help businesses to develop and adopt market-creating innovations that can create new jobs to replace those that will be lost to efficiency innovations.

If Kiwi businesses are going to learn how to adopt new technologies, and in particular new market-creating technologies, they need to step up their own skills and R&D too.[59] In the same way that going to watch Dan Carter kick drop-goals doesn't make much difference to my ability to nail a drop goal, peeking over the fence at another business using a new technology doesn't mean you suddenly know how to use it on your side of the fence. Yet business R&D investment in New Zealand is among the lowest in the OECD.[60] Similarly, businesses with higher-skilled workers and management are more likely to take on new technologies. It's a lot easier for Dan Carter to help Beauden Barrett to improve his drop kicking than for him to teach me. Yet management capability in New Zealand, some studies suggest, is not what it should be.[61]

The speed of transformation these days is so much ... the same

Is the speed of technological adoption and diffusion speeding up, as is often claimed?[62] It doesn't seem so. The McKinsey Global Institute, the think tank of the global consulting firm, looked at how quickly twenty-five technologies across the last sixty years went from commercial availability to 80 per cent adoption. Their conclusion was clear: '[T]here is no evidence that technological adoption has yet accelerated over the last 60 years'.[63] Adoption and diffusion are faster today than in the 1800s and early 1900s but they find no evidence of acceleration in the past sixty years. Comin has put together a unique dataset on technology diffusion and is widely published on the topic. He told me in an interview that taking a 200-year view, adoption lags – the time it takes for a technology to arrive at least somewhere in a new country – have definitely reduced. This is largely because poorer countries have got quicker at it. However, widespread and productive use of a new technology is much harder. This may have become quicker in some countries but his data doesn't provide a clear picture overall. However, Comin does find that some countries manage to take up technologies widely and productively more quickly than others. Much depends on each country's policies, business choices and circumstances.[64]

The McKinsey Global Institute, looking at its twenty-five technologies, finds that the time from commercial availability to 80 per cent adoption falls consistently between eight and twenty-eight years. Yet measuring from the time of commercial availability skips the critical and sometimes lengthy process of making an invention commercially viable. Diego Comin and Bart Hobijn find that for 15 technologies across 166 countries, the average adoption lag is 45 years from invention, although this varies significantly by technology and country.[65] Are these good proxies for the likely speed of adoption of the myriad new technologies that we are told will change and automate jobs? Unfortunately, both studies suffer from the same problem: they consider only technologies that were successful, such as the internet, personal computers, steam engines and mobile phones. Using the speed of adoption of some of the most successful technologies in history as a proxy for what will happen to the array of new untested technologies today is likely to vastly overestimate the speed of adoption. It's like using the rugby careers of only Dan Carter and Richie McCaw to predict how quickly each of the Manawatū under-16s rugby team will make the All Blacks.

A wide range of the new technologies discussed in studies on the future of work are only demonstrated in laboratories, or as prototypes, or simply estimated to be possible by experts.[66] Not every technology out there now that is hypothesised to impact on the future of work is going to be as successful as electricity, smartphones or the internet. Some new technologies will be complete duds and never make it to market; others will appeal to, or be affordable to, only a small niche. Think of the Segway, which had the technical capacity to eliminate the need for walking outdoors, but which remains almost exclusively the domain of overheating tourists. Or consider the infamous Hutzler 571 banana slicer – able to reduce the labour required to slice a banana – subject of endless ridiculing online reviews. Some new technologies are both useless and expensive. This doesn't mean they won't garner excitement and investment. New Zealand's Ministry of Health spent over $30,000 on Beams – basically an iPad stuck to a Segway – which have been gathering dust in a hospital basement. The Juicero, a wi-fi-connected juice presser, launched at a cost of US$699 and squeezed a proprietary pack of fruit to give a cup of juice. It quickly transpired though that you could achieve the same result squeezing the pack of fruit with your hands. Google's venture-capital arm and other investors backed Juicero to the tune of US$120 million but the nascent juicing empire has, unsurprisingly, shut down.[67]

When thinking about the future of work, starting from what we know from past technological changes and technology diffusion, and productivity today, is useful. However, we can build theories about the future that go beyond strictly reviewing the evidence. Let me suggest five rules of thumb for how technologies might diffuse and become widely and productively used in the future.

1. The more know-how needed to use a technology productively, the slower widespread and productive adoption will be.

One sunny afternoon, when I was nine, my mother taught me how to send a little plastic helicopter – a Christmas gift – whirring into the air. Sit the helicopter on the launch like this, yank the string like so, and point it away from Mum so as not to take her eye out (always the tricky part). Learning to fly a real helicopter is trickier still. It takes fifty hours of flight training and six (successfully passed) theory exams. If you think that's hard, try building a helicopter. That requires a sprawling team of designers, engineers and technicians empowered with some seriously advanced machines.

More advanced technologies normally require more know-how. So, should we expect future technologies to have greater or lesser know-how needs? The trend is towards more. Comin suggested to me that the know-how needed to use modern technologies is increasing. This, he added, could reconcile two seemingly contradictory facts. First, there is a growing divergence between advanced countries and less advanced countries in how much and how well they use new technologies. Second, this is the case despite other drivers that increase technology diffusion, like good governance and cheap transportation, converging across countries.[68] This suggests that learning to use new technologies productively is getting harder – it requires more and more know-how, which less advanced countries don't yet have. If this trend continues we would expect productive widespread use of new technologies to be more challenging and potentially even slower than in the past in all but the most sophisticated and connected cities in the world.

Yet, shouldn't a weightless computer program or application diffuse more quickly than a physical machine like a self-check-in kiosk? In other words, might the nature of some technologies make them easier to diffuse? The evidence for this so far is surprisingly weak. The McKinsey Global Institute notes that 'even highly popular and widely used social-media applications do not achieve a high level of adoption faster than technologies in previous eras.'[69] Comin similarly finds that the nature of the technology itself is far less important to explaining productive use than distance from the place it was invented. New software and phone applications like Google's new AI-powered virtual assistant, Duplex, may arrive in new countries more quickly than previous technologies but finding ways to use them productively is the real challenge. And there is little evidence to suggest the nature of technologies is making productive use easier. Instead, the extent of know-how needed to use a technology productively seems to matter more.

2. The smaller the initial change to the existing lifestyle or production that is required to use a new technology, the faster the widespread productive adoption.

Electricity revolutionised production. It completely changed the source of power for production of any sort. Goodbye steam engines, sweating humans and grunting oxen. Hello sleek copper wires. Power stations were selling electricity as a commodity in London and New York by 1882. The first electric motor was used in manufacturing in 1883. Yet initially electricity barely changed manufacturing at all. In 1900, seventeen years later, electric motors provided just 5 per cent of the mechanical drive power in American factories. It was not until the 1920s, thirty years after that, that productivity in American manufacturing took off thanks to electricity. Economic historian Paul David has an explanation: whole factories had to be reconfigured to take advantage of electricity.[70] Initially steam engines were just replaced with large electric engines with all the drive shafts, belts, secondary shafts and mechanical system still in place. Yet productivity and work changed little. To really have an impact wires had to replace drive shafts, many small motors had to replace single big systems, and factories could spread out with workers each taking responsibility for their own machine.

Making use of future technological advances may require significant reorganisations of the physical environment of work. In Washington DC beige, boxy, little delivery robots on wheels can be seen rolling along paths, currently chaperoned by a bored-looking human. Using delivery robots widely presents some challenges. What to do about the dog trying to bite the delivery robot, or the painter's buckets that have blocked the path to the front door, or the water running down the poorly drained path in a passable impression of the Huka Falls? Much like the overhaul of factories for electricity, changing the physical environment of front yards would make things smoother. Build a fence to exclude the dog, get that drainage problem fixed, and perhaps install a safe box for a delivery robot to dock onto for drop-offs. That, though, isn't exactly a seamless integration with our existing lifestyle – and the benefits seem unlikely to top electricity. In general, self-driving or self-guiding systems struggle with changeable or unpredictable environments. Even an unmapped environment is a problem: self-driving cars as they are currently designed would be completely bewildered if placed in the middle of an empty New Zealand paddock. The solution in part may be to create more controlled physical environments – but that will take time. The more complicated the changes to our existing lifestyle and production process needed to use the technology, the slower effective adoption is likely to be.

3. The price of a new technology matters as much as what it can do.

Tim Harford, the economist and writer, has convincingly shown that cheap innovations often trump magical ones. As he explains in his book *Fifty Inventions That Shaped the Modern Economy*, the American West was reshaped by the invention of barbed wire (cheap fencing). The shipping container (cheap logistics) revolutionised shipping. The digital spreadsheet (cheap arithmetic) changed office work dramatically.[71] In New Zealand it's easy to see how cost matters. Fully automated milking systems – where cows are milked with no human labour – were developed in Europe in the 1970s and have been commercially available since the 1990s. Yet, in New Zealand in 2017 there were just twenty farms using such a system; that's 0.2 per cent of our dairy farms.[72] The biggest reason for this, as farmers have said, is cost.[73] To buy the nine automated milking-system units needed to milk daily the average South Island cow herd (620 cows) would cost around $2 million.[74] That's not the kind of change Kiwi farmers can fish out from down the back of the couch.

Harford's advice for thinking clearly about future technologies: pay attention to the cheap inventions, they are the most likely to be transformational because they are the most likely to be actually used. This is the polar-opposite view to the implicit 'cost is no object assumption' that underpins many studies about the future of work. Sensors, Harford points out, are cheap and everywhere from our phones to our cars, recording information about the world. We underestimate the impact these may have on us. Machine learning, and the availability of big data, has also made prediction much cheaper.[75] This already allows Amazon to predict what we might like to buy next, Google to predict what we will type into the search bar, and even governments to predict everything from the restaurants on which they should focus their hygiene inspections, to which locations might be crime hotspots.[76] Cheap prediction should have our attention – and jobs that primarily involve prediction should expect change.

4. The smaller the number of people who have to invest in a technology relative to the number who will use it, the quicker the diffusion.

Old habits die hard. So do old preferences. Getting people to do things differently is extra difficult if there is a cost to them doing so. And getting more people to do it is harder than one or two. If a single large company, like Amazon, starts using aerial drones to deliver packages the ramp-up in drone deliveries will be extremely rapid. If, on the other hand, we had to rely on every local garden centre, dairy and boutique clothing store to each decide to buy a drone and implement a drone delivery system, the uptake would be far slower. From a New Zealand perspective the key question on a new technology's speed of diffusion may be this: will the new technology arrive through a single dominant firm sweeping our markets, or does it rely on millions of individuals making a choice to use it?

This is partly a question of market structure. Technologies that are designed for industries in which there are a few big players, like car manufacture, can arrive much more quickly than technologies that are aimed at industries that have many thousands of small players. Remember again the 99.8 per cent of New Zealand dairy farms, roughly 9,980 farms, that have not invested in automated milking systems.[77] New Zealand's large numbers of small businesses suggest adoption may be slower for many technologies.

Yet, some new technologies may help change the industry structure. Most sales of cars have been to individuals: mums, dads, 18-year-olds buying souped-up rides and so on. That's also why the transition to hybrids and electrics is very slow – it requires each individual heart and mind to be convinced enough to change what they do with their wallet. Change is also slow because people don't buy new cars that often – in New Zealand the average car on the road is fourteen years old.[78] So even if tomorrow the only car you could buy at a car dealer was self-driving (we are a long way from that), it would still take quite some time for all the cars on the road to change to be autonomous. However, while semi-autonomous cars will still be sold primarily to individuals, fully autonomous cars could generate a new transportation model that might accelerate adoption. Fully autonomous cars are more likely to be owned by corporations operating fleets that they can send at your beck and call. This could mean the number of fully autonomous cars on the road could increase more quickly than the number of semi-autonomous cars, if Waymo or Uber buys and deploys them en masse. Still, while consumers may not need to buy one themselves, they will still need to choose to use one, and potentially to choose not to buy their own car. Those usage choices will matter greatly to diffusion. This is why Elon

Musk gets so upset at the saturation coverage every time a self-driving car crashes.[79]

5. Beware the technology that affects workers in New Zealand without actually arriving here.

Diffusion is hard, and it may be particularly hard in New Zealand. But New Zealand businesses also compete in export markets with businesses from other places. If those businesses find a technology to provide the same product radically differently or more cheaply they could rapidly grab the market share of New Zealand companies – and put Kiwis out of work. Impossible Foods, started by Stanford biochemistry professor Patrick Brown, has created a plant-based burger that looks, sizzles and (in many people's view at least) tastes like meat. I've eaten a few, and they're pretty tasty, although many of my beef-eating friends are not so convinced. Lab-grown meat and milk may be the next step. As New Zealander Rosie Bosworth notes, there are already companies like Memphis Meats that have made meatballs in the laboratory. The cost of production of lab-grown meats is still far higher than the animal version, but it is coming down quickly.

Yet it can be a little hard to judge these developments, given the breathless way they are delivered. Progress in alternative foods needs to be set in context before we declare the end of New Zealand's agriculture. First, animal protein consumption has been on the rise since the 1960s. Global population is expected to grow by 50 per cent by 2050, to 9.5 billion. If by 2050 the whole population consumed animal protein at the same level as rich countries do today, that would be a 78 per cent increase in animal-protein consumption compared to today.[80] Either way, protein demand is going to go up – most likely by a lot.

Second, there have been animal-protein substitutes for a long time. I've been served soya milk, almond milk and dozens of varieties of vegeburger, as well as 'fake roast duck' and 'fake lamb chops'. Many of these are already cheaper than animal-based products, but most people don't see them as true substitutes. The central claim of new food companies for why this time will be different is that their product tastes better, or more similar to the animal version. This will likely please some vegetarians who miss animal products, but whether and how quickly it will convince non-vegetarians in large numbers remains an open question.

Third, premium products sell more than just the taste; they sell a story about cute, happy, free-range chickens, the gourmet history of steak or the creaminess of milk from Jersey cows. New Zealand has for too long remained largely in the bulk protein business, but innovations may force it to move towards more premium and commercially safer ground. That will be no easy trick at all – we have been trying for quite some time – but it could become an imperative soon. In short: beware the product that affects the future of work in New Zealand without actually being made or used here.

Where does this leave us?

Evidence from around the world, and particularly in New Zealand, tells us that two commonly held assumptions about new technologies and the future of work are wrong.

Assumption one: new technologies will sweep into New Zealand on the basis of the genius of the technology alone. As Fredrik Erixon and Björn Weigel put it in their book *The Innovation Illusion,* 'Do not expect ideas, technologies, or innovations to make it big just by showing up … no matter how brilliant they appear.'[81] Getting a new technology to be commercially competitive and supported by regulators, and to overcome incumbents to enter wide usage and actually improve productivity – that is tough. In New Zealand we regularly struggle with it.

Assumption two: there is nothing that can be done about how quickly, to what extent and for which kinds of technologies widespread adoption occurs. Yet immigration policy, diaspora policy, study-abroad programmes, business travel, exports and the FDI environment all make a big difference. Market competition, interest rates, labour taxes, R&D spending and regulation also affect how these technologies spread across the country and do, or do not, contribute to productivity. How these policy tools might best be used depends on what kind of future we want. But do we still have time to shape it? According to many we are already facing an imminent wave of technological change that will displace large swathes of the working population. Building on the lessons of technology adoption, the next chapter will take a close look at what is going on right now.

Three
Are the robots really coming?

> Our chief economic problem right now isn't that the robots are taking our jobs, it's that the robots are slacking off.
> – Tim Harford, *Financial Times*

It can feel as if there is no limit to what robots can do these days. In New Zealand a flying drone delivered pizza. In Singapore humanoid robots have been teaching kids in kindergartens.[1] Researchers at MIT have developed a device that can 'hear' your inner voice without you actually having to speak out loud.[2] And in an echo of the TV series *Westworld*, researchers are including living muscle in robots.[3]

Stories about these remarkable technologies are often accompanied by a range of studies predicting massive job losses. As the last chapter discussed, Carl Benedikt Frey and Michael Osborne's widely cited study leads the pack with a prediction (made in 2013) that 47 per cent of American jobs are at high risk of automation in the next ten to twenty years. In New Zealand a range of authors have replicated the methodology and garnered similar predictions. For the New Zealand Institute of Economic Research, writing in 2015, it was 46 per cent of New Zealand jobs that were at high risk in the next two decades.[4] Infometrics, a New Zealand economics consultancy, recently claimed that 31 per cent of jobs are at high risk in the next twenty years.[5] Accounting firm PwC has 38 per cent of US jobs and 24 per cent in New Zealand at high risk by the early 2030s.[6] Replicating an approach that defines jobs as at high risk of automation without considering whether the automation technologies are competitive or affordable outside of a laboratory does not make the approach any better. But these are some big predictions from big brand names: Oxford academics, top New Zealand research outfits and the second-biggest accountancy firm in the world.

Many of these predictions are often presented in the media with an explicit, or at least implicit, scary idea: robots are going to do so many jobs that people will not be able to find work.[7] High unemployment will become the new norm. This is why there is so much discussion about a universal basic income (UBI). The robots are going to do all the jobs, we are told, so we will all need a UBI from the state instead. Technological change has been altering what we do at work, and eliminating some jobs entirely, for centuries. But it has not yet led to widespread long-term unemployment. Nevertheless, we are told that what is distinctive about this time is that there really will be little left for humans to do. Long-term technological unemployment, in this view, will be a reality for many.[8]

The technology world's big hitters certainly subscribe to this view. Elon Musk says that 'there will be fewer and fewer jobs that robots cannot do better'. As a result, he says, a UBI is 'going to be necessary'. He worries also about what sense of purpose people will have 'if there is no need for your labour'.[9] Mark Zuckerberg is also in favour of a UBI, given 'technology and automation are eliminating many jobs'.[10] In fact, from Facebook co-founder Chris Hughes to Sam Altman, president of the famous Y Combinator incubator, Silicon Valley is stuffed with billionaires who believe automation is going to end employment as we know it.[11] Yet the incentive for individuals with a stake in Silicon Valley technology companies to talk up technological breakthroughs is pretty clear. Elon Musk seems to find a bold new technological plan every time Tesla is in trouble. As such, taking a hard look at the evidence is a must.

Many of these predictions are for the early 2030s. That is not far off. In 2030 my two-year-old niece will be only thirteen, starting secondary school. Tony Blair was leader of the British Labour Party for thirteen years; if Jacinda Ardern matched that she'd likely still be in office by

2030. According to the predictions, during her time as Prime Minister close to 46 per cent of New Zealand's jobs would have been eliminated. If almost half of all jobs in the USA and New Zealand will be eliminated by about 2030, you would expect there to be big warning signs already, and for many jobs to already be disappearing – and fast. After all, these predictions were made back in 2013 for the USA, and in 2015 for New Zealand.

If you wanted to see whether robots are putting people out of work, what would you look at? Well, if you are interested in whether people are losing their jobs to robots and winding up unemployed, the obvious place to start is the unemployment rate. (Economics isn't rocket science.) Has it been rising and hitting previously uncommon levels in recent years? If so, that might be because people are struggling to find jobs that robots aren't doing. Second, it could be that people have given up even looking for jobs as robots have made it so hard to get one. If so, we would be interested to see if the labour-force participation rate was declining. Third, we would expect occupational churn, the number of people changing occupations, to be much higher than in the past if robots were regularly putting people out of work. Finally, and very importantly, you would ask whether productivity is rising rapidly, perhaps more rapidly than ever. Productivity is a measure of how much the economy produces per hour of human work. If it were rising rapidly, that would be a sign that robots were beginning to produce things without humans needing to do very much actual work.

A looming unemployment crisis?

Being unemployed is distressing and associated with a host of mental health problems. It is understandable to be alarmed at the prospect of more of it. But are robots actually taking our jobs? They really – and I mean really – do not seem to be yet. Unemployment rates across rich countries, as I write in mid-2018, are at historic lows. In September 2018 the unemployment rate in New Zealand hit a ten-year low of 3.9 per cent, despite years of record high inward immigration.[12] In

Ashburton the unemployment rate was so low, 1.9 per cent, that they have started a promotion campaign to attract workers.[13] In late 2018, the US unemployment rate hit a forty-nine-year low.[14] In some places in the USA, businesses are having so much difficulty finding workers, they are hiring inmates even while they finish their prison sentences.[15] Over the lakes in Canada in early 2018 the unemployment rate hit a forty-year low.[16] Justin Trudeau, the current Canadian Prime Minister, was six when the unemployment rate was this low. Even in Brexit-era Britain in 2019 unemployment hit its lowest level since 1975.[17] No Brit under the age of forty-four has ever seen unemployment this low. These are unemployment lows of historic proportions. If serious technological unemployment was looming you would expect this to be the reverse; forty-year unemployment *highs* rather than lows.

Of course many factors affect the unemployment rate. And by 2019 these rates may have gone up, particularly as Brexit bites and President Trump whacks in trade tariffs that are likely to be a drag on the global economy. But it's very hard to believe that somehow every other factor is conspiring to obscure massive technologically induced unemployment. There is no escaping the evidence on unemployment; it is the opposite of what a looming robot-induced employment doomsday would look like. Moreover, even if unemployment were instead at a forty-year high, we would still have to establish that this was because of robots and technology rather than other economic problems. Maybe technological unemployment is around the corner, as some tech gurus claim. But it would be a lot more plausible to make the case it was about to arrive if unemployment rates were even just normal, rather than at historic lows.

Are robots discouraging us from even looking for work?

Unemployment statistics measure the share of the labour force who cannot find work. Yet to be considered a part of the labour force you have to be actively looking for work. Some people are not counted in the labour force for good reason: they are at school, or retired, or, like my two-year-old niece, haven't yet learnt to eat without using a high chair. Let's therefore only consider those between fifteen and sixty-four. People could be so discouraged in their search for work due to robots taking over so many tasks that they give up looking. Fewer people looking for work than there used to be could be a sign of technological unemployment.

Seen from Washington DC, the news on this appears worrying. In the USA, labour-force participation has been declining steadily since about 1997 (even earlier for men), when it was 77.4 per cent. In 2017 it was 73.3 per cent.[18] This is still relatively high by global standards. Yet as former US Secretary of the Treasury Larry Summers has pointed out, if this trend continues, by 2050 one-third of prime-age American men will be out of the labour force.[19] So far, so worrying. This has generated a great deal of concern in America and much speculation that robots (and China) are to blame.[20] There are two problems with the idea that this is evidence of impending technological unemployment worldwide. The first is that between 2015 and 2018 the labour-force participation rate in the USA for both men and women has increased again.[21] Although this is just a few years of evidence, it doesn't fit with the idea that recent technological innovations are hastening exit from the workforce. The second, and most important point, is that declining labour-force participation is a very USA-specific story.

New Zealand's labour-force participation is nothing like America's. In 2017 it was 80.9 per cent. That is the highest ever. We have literally never had more of the working-age population in work or looking for work than we do now. Growth in labour-force participation has been particularly strong for women, rising by 9 percentage points since 2000.[22] This is despite many people staying in education longer than in the past. New Zealand economics commentator Brian Fallow wonders if we have hit the 'jobs limit' in New Zealand – the point of maximum sustainable employment.[23]

This trend is not just in New Zealand. In Australia, the United Kingdom, Canada, France, Germany, Italy and Japan labour-force participation is at its highest point since the OECD started gathering data in 1960. Indeed, the same is true for the OECD on average – even when you include the USA. So, with the exception of the USA, labour-force participation rates are at all-time highs across most rich countries. That is the opposite of what we would expect if technological unemployment were just over the horizon.

If you put the data on unemployment and labour-force participation together, you get a simple equation: highest-ever labour-force participation rates plus lowest unemployment rates in decades equals more working-age people in work in most wealthy countries than ever before in history. Many countries track the employment rate – the share of the working-age population that is actually working. In New Zealand in 2018 the employment rate was the highest since records began over thirty years ago.[24] The same is true for most major developed economies.[25] It really does seem an odd time to be worrying about widespread technological unemployment.

Technology may change how much we work in other ways. As Chapters Six and Seven will discuss further, there have been some increases in part-time work across rich countries, although in New Zealand part-time work has been getting less common recently despite more people working overall than ever before.[26] Even if we assume part-time work was always involuntary, technology doesn't seem to be pushing us into it very much. Across the OECD the number of hours people work on average has declined very gradually (and very mildly) in the last thirty years, although these declines have flattened off in New Zealand and in many major economies in the last decade.[27]

Could it be that the USA is just ahead of the rest of the world in its use of technology, and that therefore we are all about to meet the same fate as the Americans with many people dropping out of the labour force? The USA is a technologically advanced country. But it is not the most widespread user of technology or robots. In South Korea robots are easy to spot. As soon as you stagger bleary-eyed off a plane at Incheon Airport there is a fleet of robots to help guide you to immigration or tell you where a much-needed toilet is. Robots have been working alongside humans at the airport since July 2017. First impressions, in this case, are not misleading. The International Federation of Robotics has a widely used measure of robot uptake in the manufacturing sector.[28] South Korea has by far the most in the world. In 2016 it had 631 industrial robots per 10,000 employees while the USA had just 189 per 10,000 employees in manufacturing. Yet labour-force participation in South Korea has been steadily rising for thirty years. Put differently, as South Korea has used ever more robots, they have also got ever more people into the labour force.

Singapore, Germany, Japan, Sweden and Denmark all also had more robots per 10,000 employees than the USA. What is happening with labour-force participation in those six countries? The highest in twenty years for all but Denmark. Even in Denmark, where labour-force participation has stayed about the same in recent years, it is at one of the highest levels in the world. As a point of comparison, New Zealand had just 49 industrial robots per 10,000 manufacturing workers – just one-twelfth of what South Korea has. The overall evidence suggests that even if we had far more industrial robots labour-force participation could remain high.

Declining labour-force participation in the USA is much more likely to be a story of social, economic and political failure rather than an inevitable result of using robots. Other countries, after all, are able to use similar technologies but generate increasing labour-force participation.[29]

As technologies change the nature of work, people adjust and adapt. This can require them to change jobs, retrain, or in some cases move from places where jobs have declined to places where there are more of them. But geographical mobility in the USA has dropped precipitously.[30] Entering the housing market in cities and suburbs where jobs are booming has become much harder. Social safety nets for those who lose their jobs are also more limited in the USA than in many rich countries. These are not the only reasons, but for countries wishing to avoid a declining labour-force participation rate, smoothing the way for people to move to where work is would be a wise step. In this light, New Zealand's dysfunctional housing market is particularly concerning.

Constantly changing jobs to keep ahead of the robots?

If we were in an age of unprecedented technological change we would expect the number of people changing occupation to be very high by historical standards. In such an age technology should be eliminating some occupations or at least vastly reducing the number of people needed in them. This in turn would require many people to enter new or different occupations. Individual stories of new technologies also suggest this should be the case. Lawyers are using automated programs to search for relevant legal cases, factories are using more and more robots, and a burger chain in California even has a robot flipping burgers. We would expect legal clerks to be looking for a different occupation, manufacturing workers to be thinking of retraining, and sous-chefs to be heading back to school.

Yet a study by Robert Atkinson and John Wu finds that the rate of occupational churn in America is at its lowest level in recorded history – right back to the 1850s.[31] That really is remarkable given, as they note, we are often told we are in an era of exponential technological change.[32] This surprising finding is also consistent with the fact that average job tenure has actually increased in the USA since 2000.[33] Similarly, in New Zealand the share of people who have been in their job for longer than five years has increased in the last decade.[34] The same is true in Australia.[35] All this is deeply at odds with the idea that new technologies are replacing workers faster than ever before.

Does productivity growth mean widespread automation is near?

'Productivity isn't everything but, in the long run, it is almost everything,' said Paul Krugman, a Nobel Prize-winning economist.[36] That's because productivity measures how effectively an economy uses its labour and capital. In the long run if workers can't produce more per hour, growth will slow. The only way to get sustained economic growth is to grow productivity. If you have better, smarter, faster machines that your workers know how to use, that should increase productivity.[37] Automation – the ultimate in better, smarter, faster technologies – should lead to rapidly improving productivity. After all, if it once took four people to do a restaurant chain's accounts, and it now only takes one using an artificial-intelligence-powered computer, that person is four times as productive as they used to be.

Productivity growth helps societies become wealthier by producing more with less. Yet it is also an indicator of potential job losses from automation; three of the four people that used to do the accounts in the above example are no longer needed. As such it is an extremely important measure of whether robots are coming. If we don't see high productivity growth, it's highly unlikely we are heading for widespread technological unemployment any time soon. Yet, as Chapter Four will explain, high productivity growth still does not mean long-term technological unemployment is inevitable – much depends on how the economy and policymakers react. In short, productivity growth is a necessary but not sufficient condition for technological unemployment.

So what's happening to productivity growth? It is at its lowest among advanced countries since the 1960s.[38] Since the early 2000s it has dropped especially precipitously. This is a source of enormous concern worldwide and the subject of countless academic studies. Many countries have reacted in a way that, while sensible, reveals how little we know about how to solve the problem: they have set up a commission. New Zealand has one, prompted by decades of low productivity growth. Australia has one. Chile too. Mexico joined the party. Colombia and the United Kingdom have one in all but name. France established a national council for productivity in 2015.

New Zealand's productivity growth has been weak for much longer than other developed economies. In fact, our lauded rock-star economy since the global financial crisis has been built on more Kiwis working, particularly the elderly and women, and on increased immigration. Our vaunted figures for GDP growth are almost always quoted in national terms, not on a per capita basis, and never per hour. That's because we aren't in fact growing our output as much in per capita terms, and we have barely been growing it at all per hour.[39]

This stubborn reality of low productivity growth alongside heightened global anxiety about automation has led to some flatly contradictory coverage. Newspapers run stories on one page saying the slowdown in productivity growth is severe and worrying – and then, overleaf, print articles claiming that automation threatens major technological unemployment. These are two mutually incompatible fears. And right now the weight of evidence is very strongly with the former.

Hype about automation and technological unemployment getting ahead of the evidence is not restricted to New Zealand. The World Economic Forum has been hammering the drum of the fourth industrial revolution since 2016, despite stubbornly low productivity growth among almost all the attendees' countries. The CEO of Google, Sundar Pichai, has said that artificial intelligence is going to be more important than fire.[40] That's quite a claim. A frustrated Tim Harford, the *Financial Times* columnist, moved to address this narrative head-on, stating bluntly that 'plenty of jobs, but lousy productivity: imagine an economy that was the exact opposite of one where the robots took over, and it would look very much like ours'.[41] That there is over-the-top hype about innovation is undeniable. Similarly undeniable is the anaemic productivity growth across the OECD. But there is also no doubt that some technological advances seem impressive. What is going on?

Forgetting what productivity growth really takes

Many economists have suggested that we are simply mismeasuring productivity growth.[42] Productivity is tricky to measure. Yet given the extent of the productivity slowdown worldwide, it's hard to see this hypothesis as anything other than clutching at straws. For one, productivity has always been hard to measure. For mismeasurement to explain the productivity growth slowdown, measurement problems would have to be worse now than they were in the past. That is not obviously the case.

The biggest concern about mismeasurement has been in the information technology sector: the possibility that mobile and online content – those YouTube cat videos and Facebook posts – are not being properly valued in official statistics.[43] However, although measurement of this sector may have initially been a challenge, we are now getting better at it – suggesting mismeasurement may be becoming less of a problem, not more. Moreover, the productivity slowdown has occurred across many countries and is not related to the size of their information technology sectors.

Finally, for mismeasurement to explain much of the slowdown we would have to be measuring it spectacularly badly.[44] When you need an enormous and unprecedented measurement error (for which there is little evidence) to explain away the decline in productivity growth, you are pursuing the wrong theory. As Daron Acemoglu told me, 'Mismeasurement is really not an explanation for what we are seeing.'[45]

The global productivity-growth slowdown is real. Perhaps amid the hype over robots and artificial intelligence we've forgotten how momentous previous eras of innovation were. Above all maybe we've forgotten how transformational a technology needs to be to have a large impact on productivity growth – and our current technologies simply are not as remarkable as we sometimes think they are. Robert Gordon, an economist at the University of Chicago, is a leading proponent of this view. He argues that the low-hanging fruit of impactful innovations have already been picked. In his 750-page tome *The Rise and Fall of American Growth,* Gordon painstakingly lays out the evidence that earlier innovations were truly extraordinary and that the slow productivity growth we've seen since the 1970s will continue.[46]

The period between 1920 and 1970 was a remarkable time of technological change. Consider a boy in New Zealand in the 1920s; call him Tom.[47] He would have been accustomed to getting around by walking – or saddling up a horse. By the time he was a pensioner in the 1970s, horses were for racing. Cars were totally dominant in transportation. More dramatically still, by then people thought relatively little of literally *flying* across the country for a work meeting. Unless Tom was born with a silver spoon in his mouth, electricity for lighting and heating would have been a distant dream in his early years. If he was lucky, he might have had indoor plumbing. As a pensioner Tom and every friend of his would think nothing of drinking water out of the kitchen tap and flicking on a reading light while warming their feet with an electric heater. Flush toilets were the envied privilege of the rich few when Tom was born. By the time he was sixty it would have been unthinkable not to have one.

Radio was invented in his lifetime, followed by black-and-white television, and by the time Tom was in his sixties he could watch the 1972 All Blacks test against Australia live and in his own home. At half-time he could get a cold beer out of a refrigerator – a device that hadn't been invented when he was born. If he sloshed beer on his shirt celebrating a try, he could put it in a washing machine and then chuck it in the dryer. None of this was possible when he was born.

Infant mortality, a real threat to Tom in his early years, was extremely unlikely for his grandkids. They were also expected to live a decade longer than he had been expected to. Transport Tom back from the 1970s to his childhood home in the 1920s and the lack of technology would have been a heck of a shock. On the other hand, transport a fifty-year-old Kiwi today back to their childhood home in the 1970s and they would find a grainier television with fewer channels, likely no microwave, and of course a missing smartphone. But they would probably miss relatively little – and certainly not have to struggle with chamber pots and hand-washing clothes.

Not only was the 1920s to 1970s an extraordinary time of innovation, in Gordon's view we are unlikely to see anything like it again. His reason for this is simple: certain kinds of fundamental progress can only happen once. Access to indoor electricity, indoor plumbing, almost eliminating infant mortality, the ability to communicate directly with someone who isn't physically present: these were hugely transformational and cannot be reinvented. The big, juicy technological fruits, he claims, have already been plucked. Gordon happily concedes that innovation will continue and will at times seem very impressive. But will sexy innovations like turning down the lights using Amazon's Alexa transform life in the way that wired electricity in the house did? You can watch a lot more channels on your TV than you used to be able to, but is that as transformational as having a TV in the first place? Gordon's answer to these questions, and many others like them, is absolutely not.

Gordon may just lack a sufficiently vivid imagination for possible future technological innovations. Nonetheless, he is convincing in arguing that the bar for an invention to have an impact on productivity is very high. Consequently, he is lukewarm on the prospects for 3D printing, big data and artificial intelligence, and driverless cars to significantly impact productivity. Making a car driverless is nothing compared to inventing the car in the first place, he argues, and drivers can already take calls and listen to radio news and music as they drive. The marginal productivity benefit of also being able to look at a computer screen as you commute is likely to be small. The big-impact technologies, in his view, are already invented.

Are the robots just late to the party?

There is a simple opposing view to all this: just wait. This is the lull before the storm. This view is worth taking seriously. New technologies can take quite some time to show up in the productivity statistics. Michael Mandel and Bret Swanson, two American think-tank economists, made their views clear in the title of their 2017 report, *The Coming Productivity Boom*.[48] They claim the productivity drought is almost over. The main driver in their view will be the arrival of the information-technology revolution in physical industries. They point out that productivity in the digital industries in the USA grew by 2.7 per cent annually over the last fifteen years but in the physical industries it increased just 0.7

per cent per year.[49] Their central claim: the productivity performance of digital industries can be translated into other industries thanks to better investments and advances like the internet of things. So far we are still waiting.

The broader riposte is that development of computers, information technology, email and the internet in recent years made a small impact on productivity statistics compared to the period between 1920 and 1970. Moreover, this impact seems to have tailed off almost completely, even though most physical industries use all of these.[50] The transition from paper to electronic catalogues, Gordon notes, happened fifteen to twenty-five years ago and produced a one-time jump in the level of productivity but no permanent increase in its growth rate.

Nevertheless, there are many new technologies under development that appear as if they could take over tasks that have been typically done by humans. A New Zealand company, Soul Machines, has developed a digital assistant to help customers deal with queries online. They claim that the digital assistant shows both enough artificial intelligence and emotional intelligence to chat with customers about their queries. ANZ Bank and Air New Zealand have both trialled the technology, with the aim of the digital assistant taking over some of the low-value conversations people have with real staff. Similarly, digital assistants like Alexa are being trialled to take over some requests in hotels, including the Marriott chain, potentially replacing concierge services.[51] In offices, artificial-intelligence bots are being trialled to automate basic office tasks like accounting, billing and payments. These bots are already saving on tens of thousands of work hours that used to be done by people.[52] Retail stores like Amazon Go are being developed, where there are no cashiers and shoppers simply walk out with an advanced sensing system detecting what they have taken with them and charging their digital account. Manual tasks may be affected; robots on production lines are widespread and now at Auckland airport there are three robots used for cleaning.[53]

Another source of disruption could be what the economist Richard

Baldwin calls 'globotics'.[54] New technologies could make it much easier for workers outside New Zealand to provide services here. Imagine consulting a lawyer in Guangzhou about your boundary dispute in Tauranga (aided by real-time artificial-intelligence language translation), or 'going to' a medical check-up with a doctor who is connected to your clinic in Whāngārei via video or hologram from Delhi. Yet even if the technology allows this, much will depend on how much we value and benefit from in person meetings and care. As we saw in the last chapter, despite the hype and advances in video-meeting technology, in-person work is still far preferred – spending on business travel is growing twice as fast as global GDP.

Still, some occupations that have recently seen a surge in hiring may not see this surge last long. With the growth of e-commerce there has been a rise of jobs for humans in warehouses, picking goods off shelves and readying them to send out. The key part of this process that requires humans is picking products off shelves. Yet an array of companies are working to develop robot hands that can reliably pick up different-shaped objects, potentially displacing human workers.[55] Many analysts see these developments and believe an automation surge must be imminent with worrying implications for workers.

Erik Brynjolfsson and Andrew McAfee, authors of the influential *The Second Machine Age,* are leading proponents of the idea that extraordinary advances of machines and artificial intelligence mean a productivity and automation boom is just around the corner.[56] Brynjolfsson, for example, disputes Gordon's scepticism about self-driving cars. The USA currently has about 3.5 million people working as drivers. If fully autonomous vehicles became commonly used, Brynjolfsson suggests the same output could be achieved with perhaps two million fewer workers. That would be a major upheaval but could boost productivity growth in the USA by 0.17 per cent per year if the phase-in came over ten years.[57] New Zealand has about 50,000 full- and part-time drivers.[58] If there were a similar reduction in New Zealand that would mean 28,500 fewer people working as drivers without any loss of economic output. This would be a massive change to the sector, but would certainly increase New Zealand's productivity growth.

Call-centre workers could also be replaced by artificial intelligence, Brynjolfsson suggests. In the USA about 2.2 million people work in call centres. In New Zealand in 2013 about 10,000 did.[59] If artificial intelligence handled 60 per cent of those calls, Brynjolfsson estimates US productivity could increase by 1 per cent, spread over ten years. Again, a similar trend in New Zealand could see 6,000 fewer jobs in call centres and a small increase in productivity growth.

Yet some see these shifts as less a case of 'just wait' and more a case of 'just wait ... a fair while'. New Zealand's Ministry of Business, Innovation and Employment still rates the job prospects for drivers as good, projecting increases in the number of truck drivers through to 2026. The ministry does forecast a gradual decline in the job prospects of contact-centre workers, but suggests there will still be about 8,000 jobs in the sector in 2026.[60] A 2018 survey of business in New Zealand found sales representatives, including contact-centre workers, as the second most in demand skill in the country, with drivers also in the top ten.[61]

Brynjolfsson does highlight the paradox: transformative technologies seem to be everywhere except in the productivity statistics. Yet he and his co-authors argue there is no contradiction between forward-looking technological optimism and backward-looking disappointment.[62] Productivity growth in the very recent past correlates with very-near-future productivity growth, but according to Brynjolfsson the productivity performance of the last ten years does not help much in predicting the productivity performance of the next ten. So despite a decade of weak productivity growth an improvement could be around the corner. To be clear, it also might not be; the data does not suggest that weak productivity performance today is predictive of strong productivity performance in the future either.

Brynjolfsson's point is well made, but he might not be making it at all if productivity growth were already high. He'd be writing about how the high productivity growth rate confirms his hypothesis about the dramatic advances of technology. Indeed, as his frequent co-author McAfee concedes, 'low productivity growth does slide in the face of the story we tell about amazing technological progress'.[63] The paradox, after all, is the subject of his study because it isn't what we would expect. In short, the poor productivity growth of recent periods does not rule out a robot-induced productivity boom soon. But it does tell us it definitely is not here yet – and strongly suggests it is not exactly banging at the door.

Even if productivity did pick up, it has a long way to go to reach the levels seen between the 1920s and 1970s. During that period productivity growth in major economies was regularly above 4 per cent, and sometimes touched 6 per cent. Today it is closer to 1 per cent in New Zealand and elsewhere.[64] To reach 4 per cent in the next decade would require the kind of sharp uptick in productivity that hasn't been seen since the immediate post-Second World War economic boom. Indeed, the McKinsey Global Institute recently suggested that in the next decade, if governments took all their advice (and as a former McKinsey consultant I can tell you that is an almost comically optimistic scenario), productivity growth could reach 2 per cent (this also assumes of course all the advice is spot on).[65] That's still below recorded productivity growth in the early 2000s. In other words, even if we do see a serious improvement in productivity it is highly unlikely to be historically unprecedented. That is important given we have had periods of much higher productivity growth in the past without ending up with widespread long-term technological unemployment. Remember that higher productivity growth is a necessary but not sufficient condition for technological unemployment.

Robots are not selling like hot cakes

Beyond the basic question of how impressive new technologies really are compared to past breakthroughs, there are other reasons why productivity growth is anaemic. In the last chapter we saw how important the adoption of technologies can be to productivity in New Zealand. Those adoption challenges are also extremely important in other places. Businesses in Boise, Idaho, or Carlisle in the United Kingdom can be slow to adopt new technologies and incorporate them productively into their business practices, just like businesses in Timaru. The International Monetary Fund (IMF) finds that investment in new capital – actually buying the damn robots and machines – has been very weak in the last decade.[66] Robots may be able to help increase productivity, but they won't do so if they are left on the shelf. The fact that businesses around the world aren't pouring money into robots also tells you something else: business doesn't think the rewards of automation are that big.

Productivity growth also depends on who is using the robots. If almost all the robots are used in a niche sector that is a tiny part of the economy, the impact on productivity will be low. To take an odd but instructive example, New Zealand could have a hyper-productive way of providing babies with their standard vaccines, but that is not going to make much difference to productivity for the economy as a whole. Providing vaccines is a tiny part of overall economic activity, and it basically only grows if more babies are born each year. If the robots in question are only good at giving babies jabs, then their impact on future productivity growth would be limited and the displacement of workers would be very limited. But if the robots are 'general purpose' and could also help in other larger sectors, then it may just be a case of waiting for the 'big' sectors to catch up with our (oddly) tech-savvy vaccinators. That would be a reason to think productivity growth might pick up soon.

Productivity growth in the whole economy does not just depend on big sectors using the robots too and becoming more productive. It also depends on whether those productive sectors use more of the economy's resources and therefore increase their share of the economy.[67] If highly productive sectors instead become a smaller part of the economy, national productivity growth could stay weak. Could this be what is holding productivity growth back? Perhaps one day.[68] But for now there just is not much evidence of strong productivity growth in any major sector of the economy in rich countries. Consider six major sectors in Europe and the USA: automotive, finance and insurance, retail and wholesale trade, technology, and tourism and utilities. Comparing 2010–14 to 2000–4, productivity growth was on average significantly lower in all areas but tourism (the lowest to begin with), in which it flatlined.[69]

There is some evidence that leading companies within certain sectors have stronger productivity growth. As we saw in the last chapter, the issue is that the technologies supporting these advances are not diffusing. The OECD bluntly concludes that while some leading firms are growing their productivity well, the technology diffusion machine is broken.[70] The first step is for the productivity growth of these few leading companies to be reflected in the rest of the sector, then in the wider economy, and then in other countries.

Higher productivity growth certainly does not mean long-term technological unemployment is inevitable. With the right policies, we have seen high productivity growth without long-term technological unemployment in the past. But if we are going to see a growing number of people permanently out of work because of technology, as many people fear, we would have to also see higher productivity growth in at least some sectors. And we just do not see that. Moreover, economy-wide high productivity growth coupled with notably high and rising unemployment would give the most credence to the claim that we are running out of work for humans. We do not see either of these trends in New Zealand or other rich countries. Indeed, we see the opposite.

Why New Zealand may be less susceptible to automation

Robots and artificial intelligence won't be brought into every bit of the economy with the same speed, nor with the same impact. Artificial intelligence may yet prove to be a general-purpose technology, like electricity. But even electricity was more transformative in some sectors, such as manufacturing in factories, than in other sectors, such as agriculture. Nor did it find useful applications equally quickly in every industry. Similarly, technologies for automation are more advanced in the manufacturing sector.[71] This is important to New Zealand's prospects. Our economy is quite distinct from many other rich countries. We spend more time in fields, less time in factories, and a lot more time showing other people our glaciers, boiling mud and beaches. To see how our predilection for farming and tourism alters the picture on automation for us, we need to take a closer look at exactly what robots are likely to be able to do.

OECD researchers have completed two fine-grained studies of automation's potential in which they break occupations down into specific tasks and ask which of these tasks could be done by a machine. By contrast many other studies, most notably Frey and Osborne's, assume that whole occupations are automated. If some of the many tasks that make up a given occupation aren't in fact automatable, this approach risks overestimating how many jobs could be automated. My uncle-in-law drives a school bus. Moving the steering wheel, shifting the gear stick and indicating left and right are central to his job. But among other things he also makes sure Peter doesn't punch John, that the kids sit down while the bus is moving, and that if someone is unwell they get the attention they need. Self-driving buses may one day be possible, but someone also needs to supervise the kids. A more refined approach that focuses on tasks can give better insights and much-changed answers.

Using this more refined approach, one OECD study estimates that 9 per cent of jobs in the OECD are 'potentially automatable'.[72] That's a lot less than the oft-quoted 47 per cent of jobs at high risk of automation. Another OECD study extends the analysis to thirty-two countries, including New Zealand.[73] Its estimates are slightly higher for rich countries on average, with 14 per cent of jobs at high risk of automation. But it also finds that the average Kiwi's risk of having their job automated away is lower than in any other country. This is in part because a lot of Kiwis work in areas like professional services that are difficult to automate, while fewer work in manufacturing, which is easier to automate. Even these estimates, which give much less alarming predictions, assume that everything that technically could be automated will be. As Chapter Two showed, this assumption is very difficult to defend.

The fact that New Zealand is likely to be substantially less susceptible to automation than other countries is on the one hand reassuring but on the other hand problematic. Automation helps raise productivity, and that in turn allows the economy to produce more and supports wage and income growth for New Zealanders. Avoiding dramatic job losses is reassuring, but if we miss out too much on the benefits of automation we may find ourselves falling further behind other rich countries in income terms. This is why the right mix of types of innovation really matters if we want to maintain a world of growing incomes and full employment. Efficiency innovations like automation help make us wealthier but also eliminate jobs. As we will see in the next chapter, for growing both wealth and jobs, automation needs to be complemented with market-creating innovation that generates new tasks for humans to do.

Unprecedented change is unlikely

One question that is rarely asked in studies of automation and job loss is how big is big? How many job losses is a lot? Nine to fourteen per cent of jobs being eliminated still sounds like a lot. But context matters; in the course of a regular year in New Zealand approximately 15 per cent of the workforce loses their job.[74] The vast majority of them find a new job. Changing job can of course be stressful and very difficult. But seen in light of what people and the economy do every year, 9 to 14 per cent due to artificial intelligence does not sound quite so alarming. If job changes due to automation are in addition to these regular job changes, and are a result of the elimination of entire occupations, then of course the challenge is greater. Still, over the next forty years more than ten million jobs will be lost in New Zealand due to normal market turnover. As the Artificial Intelligence Forum points out, even in a bad-case scenario in which a million jobs are lost to AI in New Zealand, that would only be 10 per cent of the change in jobs in that period.

It is unlikely that the next fifteen years will see unprecedented shifts in employment. We in New Zealand, and in other developed economies around the world, have come through bigger changes to employment than this. The McKinsey Global Institute has made its own predictions about job losses due to automation. These are task-based rather than at the cruder occupational level, and suggest that while a range of tasks could be automated, only 5 per cent of occupations are fully automatable based on current technologies. They then attempt to account for the process of technology adoption through to 2030, including the adoption of technologies that automate tasks even if not the whole occupation, and estimate the impact on employment. They compare these to historical episodes of major employment change around the world. The headline news: we've seen bigger shifts before. They suggest that 'future sector declines from automation are largely expected to be within range of historical declines on a sector basis, but smaller as a share of the overall economy'.[75] For example, the share of workers displaced in transportation in Germany in the next fifteen years is expected to be smaller than the share of workers that moved out of German agriculture between 1957 and 1971, when half the agricultural workforce transitioned out. No single prediction of future automation in any sector anywhere in the world is expected to cause a larger decline in employment than the change in manufacturing that the USA went through in the 1990s. While major workforce transitions like this can be very tough, the 1990s are rightly remembered as a time of economic strength in the USA. They were, after all, the heart of the longest recorded economic expansion in American history.

Uncertainty remains

This is not to say that we have nothing to worry about. Change to work, even if the vast majority of jobs aren't eliminated, could be very disruptive. And if dire warnings of job losses do come true, even if

New Zealand is less severely affected, there is little reason to celebrate being the one-eyed man in the kingdom of the blind. This is especially true if we also miss out on potential productivity gains that could raise incomes. Nevertheless, we should not take it as a given that the story elsewhere in the world on automation will be New Zealand's story too. Thinking clearly about our options for the future requires a realistic grip on the challenges ahead. The evidence thus far suggests that the hype is quite a long way ahead of the reality when it comes to job losses and automation. Unemployment is at record lows, labour-force participation at record highs, occupational churn at record lows, and productivity growth at record lows. These are the opposite of what we would expect if widespread technological unemployment was near.

Of course in the long run things could change or accelerate in surprising ways. Digital assistants, shops without cashiers, fully automated e-commerce warehouses, robot cleaners, self-driving cars, automated call centres and digital concierge services could all displace human workers from traditional tasks. Much will depend on the speed and extent of commercial development and adoption. As Chapter Two has suggested, this is often slower than we anticipate. Still, what if we did see rapidly growing productivity, and rapid development and adoption of technologies that take over many previously human tasks? Would New Zealanders still have jobs? The next chapter will attempt to provide some answers.

Four
So they just take our jobs and that's it?

Now is the time to understand more, so that we may fear less.
– Marie Skłodowska Curie, quoted in Benarde, *Our Precious Habitat*

There have been so many alarming headlines: 'Will machines devour man?', 'Machines as ministers to man', 'Does machine replace men in the long term?', 'Machines are laughing at men', 'Workers see "robot revolution" depriving them of jobs', 'Automation might end most unskilled jobs in 10 years', 'Labour "slag heap" feared by Writz', 'PM acts to counter "chips" that kill jobs', 'A robot is after your job', 'Technology called threat to 5m jobs', 'Computers: worker menace?', 'Unemployment will end honeymoon with robots'.[1] Amidst all this media worry the White House launched a 'blue ribbon' Commission on Technology, Automation, and Economic Progress.[2]

I have not accidentally copied my introduction from Chapter One because I've run out of snappy starts. These are all real headlines from publications like the *New York Times*. The difference is that the first is from 1921 and the last from the 1980s – thirty years ago. The blue ribbon commission was launched in 1964.

At one point or other, pretty much every big hitter has been worried about robots taking jobs. Albert Einstein blamed machines for the joblessness of the 1930s in a speech in Berlin in 1931. Henry Ford was the author of the piece headlined 'Machines as ministers to man'. Another of the above articles was written in 1949 by Pulitzer Prize-winning journalist Hal Boyle. John Maynard Keynes feared we were being afflicted by 'technological unemployment'. The term caught on. Whenever unemployment spiked, use of the term 'technological unemployment' in the *New York Times* spiked with it.[3]

There have been periods of very rapid technological improvement since 1920, leading to big changes in what people work on. As the last chapter showed, if you were transported back to an average house in New Zealand in the 1920s you would miss all manner of technologies, from cars and planes to electricity and flush toilets. There have also been difficult and complicated economic times in many countries through the twentieth century. But none of these fears of sustained widespread technological unemployment have so far been borne out. This is at some level puzzling. How can it be that technological progress and productivity growth can surge ahead, eliminating the need for humans to do various tasks, and yet employment can stay high in the long run? This chapter will attempt to explain. Further, it will ask if the incentives and conditions that have allowed this in the past might be disappearing.

A workhorse model of technological change and the economy

Weak. That's probably the most truthful description of the evidence so far that recent technological breakthroughs are substantially raising national productivity, automating people permanently out of work and raising long-term unemployment. But the future could be different. I could be wrong about both how effective these technologies might be and, more importantly, how quickly they might enter popular and productive usage. Let's say I am. Let's say instead that these innovations are as good as their inventors in Silicon Valley say they are (let's not say AI is better than fire though). And let's say that people and businesses the world over, and especially in New Zealand, embrace them with the enthusiasm and speed of office interns rushing toward a free lunch. In this scenario you would be able to see the robots everywhere and a lot of automation. What is likely to happen to employment?

Confronted with a scenario like this, a basic workhorse model grounded in economic theory can help cut through the confusion. This model breaks down different effects on workers when technology progresses. Some are good, some not so good. Critically, the overall impact on workers in practice depends on the balance between these different effects. Looking carefully at how and why these different effects happen can help us think clearly about what might happen to work in the future as new technologies arrive and automation advances.[4] It can also help us see what really matters to keeping people in work.

First, if robots are able to do some tasks people currently do, fewer people will be needed to do those tasks. This will reduce the demand for workers and possibly reduce their wages.[5] This is what we hear a lot about. Let's call this the displacement effect. The displacement effect is bad for workers.

Robots replacing humans in some tasks will also raise productivity, doing more with less, because it will allow those tasks to be completed more cheaply. Producing more cheaply over time means cheaper goods and services. And lower prices mean people can afford to buy more, even with their existing incomes.[6] It's not Oprah Winfrey saying, 'You get a car, you get a car ... everybody gets a car' to the whole audience of her TV show.[7] But consumers do get the *equivalent* of a pay rise. We can spend the cash we save on new things and experiences. If we spend some of this on goods from the same industry that is using the robots, as long as there are *some* workers still needed in that industry, it will help increase demand for workers. Let's call this the home-sector productivity effect. This impact can be surprisingly strong, as we will see, although it is not inevitable. Similarly, if we spend the cash on experiences and goods from sectors that have not experienced a technological boost it will increase the demand for workers in those industries.[8] Let's call this the other-sector productivity effect.[9] Both productivity effects are good for workers.

Businesses, just like iPhone fans, like to have the latest model of their gadgets. The robots and machines that take over some tasks from workers will therefore also get upgraded over time. The newer versions will do the task even more effectively and cheaply. Yet this doesn't displace workers nearly as much as the original installation did – they've already lost their jobs to the first installation. But the new machines will further reduce the cost of doing the task, and therefore reduce the cost of the product or experience produced. This therefore doubles down on the two productivity effects (a step closer to Oprah levels of generosity). Everything is cheaper still, consumers are wealthier, and they spend this on products and experiences that, of course, require some workers to produce. Let's call this the upgrading effect. The upgrading effect is also good for workers.

A lot of analysis of the impact of technological innovation on workers stops with the displacement effect. Workers lose their jobs, end of story. The productivity effect and upgrading effects are ignored and consequently the headlines are pretty alarming. But in fact the story is still not over. In times of rapid technological change and automation there is also another effect: the invention of completely new tasks. Let's call this the new-task effect.

The process of automation itself can make the creation of new tasks more likely. This is because automation reduces the demand for workers and can allow business owners to capture more of the profits at the expense of workers. That makes further automation less profitable as workers are easier to find and, particularly among those displaced, may not be able to demand such high wages. There is therefore less profit to be made replacing them. It also means there are more people looking for work. Coming up with new tasks that available workers can do is therefore quite profitable, for both businesses and workers.[10]

Additionally, new technologies themselves can also open up new tasks for people. For example, the invention of the automobile created the job of car mechanic and countless others. Similarly, the introduction of artificial intelligence will likely bring a host of jobs such as trainers of artificial intelligence systems. New-task creation is a big part of what market-creating innovation should deliver – and why governments should care about it so much. The new-task effect is good news for workers.[11]

That's the workhorse model grounded in economic theory. The displacement effect reduces the demand for workers. The productivity, upgrading and new-task effects all work in the other direction. Market-creating innovation in particular can help generate new tasks. Policy makes a big difference to this, but there has also typically been a natural tendency toward balance between the effects. When automation gets ahead, the workings of the labour market support a rebalancing towards workers via the creation of new tasks. When workers get ahead, the incentives support a shift towards greater automation.

This is a good approximation of the broad history of economic growth. We haven't had sustained massive unemployment or big drops in wages (quite the opposite in fact), despite extraordinary technological change since the early 1800s, because these effects have roughly balanced against each other. Instead the story of modern economic history is: we produce more efficiently, we consume (new) things more significantly, and we continue to employ widely. The central claim of those who believe the future of work is one of significant joblessness is that this relationship has broken down.

Displacement and the home-sector productivity effect

Theory is useful, but do each of these effects really play out in real life? The displacement effect is easy to spot because it often makes the news. In the automobile-manufacture sector there has been lots of it. Long ago, the cars at your local car yard in New Zealand were made using a lot of human hands and hours. But even by 1961 General Motors was using industrial robots to do spot welding. This meant fewer human hours spent welding for each car. Automation in car manufacturing has since reached far more parts of the production process. For workers who lost their jobs this has been a painful process.

In New Zealand the displacement effect is easy to spot too. Wander through Canvastown in Marlborough, or read Eleanor Catton's masterpiece *The Luminaries*, set in Hokitika in the gold rush, and you get a sense of the sheer labour intensity of gold mining in the old days. It's hard not to sympathise with Catton's character Nilssen when he describes the 'miserable work – lugging pails of water to and from the river to sluice the stones, slapping at the sandflies that crept up his jacket until he was mad enough to dance'. All this for a 'pinch of grit ... yielding, at last, five dirty shillings'.[12] Today we still have the same objective: find gold. But technology means we do it differently. Today's massive underground gold mine at Waihī is a dramatically less labour-intensive operation. Instead it relies on explosives, dump trucks, crushing machines, mechanical grinders, elution and electrowinning, a furnace and a lot of water pumping. Tellingly, the illustration of the nine-step mining process on the Waihī mines' website barely includes a human.[13] Workers, without a doubt, have been significantly displaced.

Yet using more machines and robots in a sector can actually *increase* employment in that same sector, thanks to the home-sector productivity effect. This is counter-intuitive, but real-world examples abound. In the late eighteenth and early nineteenth centuries weaving cloth was big business in England. Initially it was highly labour-intensive. In 1784 Edmund Cartwright invented the first version of the power loom. It took decades for this to be perfected (commercialisation and adoption takes time), but eventually power looms automated 98 per cent of the labour needed to weave a yard of cloth.[14] That's a massive displacement effect. Yet the lower labour cost of production per yard of cloth meant lower prices. Most people at that time owned only one or two sets of clothes. Suddenly, many people could afford to change their underwear more often (or not wash their clothes so often). Thanks to the cheaper prices, demand for cloth shot up. This meant that, despite 98 per cent of the task of weaving cloth being automated through the 1800s (and about 75 per cent of it by as early as the late 1820s), employment kept growing sharply in the cotton textile sector until the 1920s.[15] That's the productivity effect – massive again.

James Bessen, an academic at the University of Boston to whom I spoke for this book, is perhaps the world's most keen-eyed observer of the productivity effect. He points out that few occupations have been as directly targeted by automation as bank tellers. ATM literally stands for automated teller machine. Automation doesn't get much more explicit than that. Bank tellers hearing about this new invention in the 1970s must have been very worried. Indeed, in the USA the introduction of the ATM reduced the number of tellers at each branch of banks in urban areas from twenty to thirteen. There's the displacement effect. But this meant each branch was cheaper to operate. Now big banking chains could reach more customers by opening more branches, in part to help with their marketing. They did this to such an extent that human bank-teller jobs in the USA continued to increase. Moreover, since 2000 they have been increasing at a rate faster than the labour force as a whole. However, as we will see below, this kind of effect does not last forever.

There are many examples of this kind of productivity effect.[16] When scanning technology was put into cashier systems, reducing the amount of labour needed to check out a person at a supermarket, the number of human cashiers increased. When lawyers gear up for a lawsuit they need to get hold of all the documents that could be relevant for the lawsuit. This laborious work was typically done by paralegals – like Meghan Markle's character in the hit legal TV series *Suits*. In the 1990s electronic software was introduced to do this search process. Yet the number of paralegals increased rather than decreased; lawyers and judges decided to do more frequent and extensive searches now that it was cheaper and, because not every task in legal discovery was automated, more paralegals were required in total. The overall picture backs up the cases: Bessen finds that jobs have grown faster in occupations that use computers than in the overall labour force. This is even true for occupations that involve more routine tasks; if they use computers, jobs have grown faster.

These findings surprised me, but they are hard to argue with. So what's really going on here? First, most jobs require the completion of a long series of interrelated tasks. It is very difficult to automate absolutely every task that makes up a particular job. Bessen finds that of the 271 occupations listed in the US census since 1950, only one has been entirely eliminated by automation: elevator operators.[17] Very few occupations are likely to be completely eliminated by automation in the next decade.[18] Technology may automate many tasks in the chain required to make a product or service. This reduces the amount of human work needed to produce a given product or service. On the other hand, this automation should also make the product or service cheaper and therefore increase demand for it, meaning many more products are made.[19] As long as workers are still required for some tasks in the production chain, more demand will also be a force to hire more workers. If the increase in demand for the product is sufficiently large, it can more than offset the reduction in labour needed to produce each product, leading to an overall increase in jobs in the sector. This is what happened in cloth production and with paralegals.[20]

The remaining human workers in the production chain are also often made more valuable, for the simple reason that without them the whole production process would fail. David Autor, an MIT economics professor, calls this the O-ring principle, after a tragic NASA mission in 1986. The shuttle launch initially appeared to go as expected, but seventy-three seconds into the flight it broke apart, killing the seven people on board in front of a horrified live television audience. The catastrophe happened because the rubber O-rings on the solid rocket booster did not seal properly due to the abnormally cold weather that day in Florida. The O-ring was critical to the space disaster because everything else worked perfectly. In many jobs, humans are the O-ring. Even if we do only a few tasks, without us the rest of the process fails. Our importance can also help our pay prospects. For example, the 2 per cent of the process of making cloth in England that wasn't automated and that workers still did became more valuable. Indeed, wages for weavers rose sharply as the power loom was introduced.[21]

Even where automation appears to be almost total, such as with ATMs, the nature of the 'automated' job may change to take advantage of new opportunities. This can keep employment growing in the same sector that is hit by automation. Bank tellers began to do less money-handling and more marketing of banking products and building relationships with customers. Thought of differently, tellers expanded their core tasks and made themselves into critical O-rings in a slightly broader production process: selling more complex financial services to customers.

Sometimes automating a task, and the associated reduction in the cost of doing that task, may reduce employment in one occupation but stimulate hiring in closely related fields. After all, any worker who benefits from the new technology is more productive and therefore better value for money. For example, voicemail systems and computerised phone lines meant fewer jobs for telephone operators, but this was significantly outweighed by an increase in the number of (newly more productive) receptionists and secretaries.[22] Similarly, there are fewer typographers today thanks to computers, but many more graphic designers and desktop publishers.

Yet in cases where there is an increase in employment in the sector that is hit by automation, it does not last forever.[23] In cloth and clothing production, at some point there is a limit to how many clothes people want, even at cheap prices.[24] And while it turns out that the limit is an extraordinarily long way from one set of clothes – people from the 1880s would be shocked by how many clothes we have – it does exist. At that point further reductions in the price of a product do not increase demand. Now new technologies can still reduce the need for workers but consumer demand does not compensate enough, so fewer workers are needed overall. This happened in the textile and cotton sector in about 1920, and employment in that sector began to decline sharply. In New Zealand, a similar dynamic may be happening with bank tellers, as customer demand for a nearby teller ceases to outweigh further automation of individual banking tasks through online banking and smart ATMs.

Strikingly, however, some of today's most advanced robots are not yet consistently reducing employment even in the sector they are deployed in. Industrial robots are amongst the most advanced forms of automation currently available. These robots do not need a human operator, and can be reprogrammed to various different tasks like welding, painting and assembling. Their impact on employment, however, seems to be mixed. One American study suggested that robots are to blame for some reductions in employment in the American Midwest. This was alarming, but also controversial.[25] Using the exact same measure of robots but looking at Europe, another study found that using industrial robots did not reduce overall employment and helped increase both productivity and wages.[26] It also found that in Europe the prices of goods produced in the sectors using robots declined, potentially meaning people bought more of them.[27] These differences suggest that the domestic demand and policy context really matter to how businesses react to automation opportunities. If New Zealand can get its policies right, further industrial robots could, for a time at least, be good news for wages and productivity without hurting employment overall.

If policymakers want to know how the adoption of a new technology is going to affect jobs in an industry, understanding the technology alone is not enough. These other factors matter too; in particular, demand matters. How much more of this product might people buy if the price drops? A big increase in what they buy would see employment increase overall in the sector despite a reduction in the hours of work needed to produce one bag of milk powder, bottle of wine or animation in a movie. On the other hand, if there is not much extra demand for the product when the price drops, employment may instead immediately begin to fall. Second, how long can that high demand be sustained? At some point consumers will not want more, despite cheaper prices, after which any further automation will see employment decline. One of the challenges for policymakers in New Zealand, if we do see rapid productivity growth, is how to keep demand high for the things we produce. The next chapter will consider this in New Zealand.

Getting jobs elsewhere thanks to the other-sector productivity effect

After peak demand in an industry is reached and employment begins to decline, does that mean we should expect people to slip into long-term unemployment? Not necessarily: the second type of productivity effect, increased spending in other sectors, is one reason why. As automation means that whatever the robots are producing will be cheaper, we would be able to afford to buy more stuff even without any increase in what we receive each pay day. Economists call this an increase in our real income.

If I had been around in the 1950s, I could have spent my cash boost on more things from the highly productive sector, in that case clothing, or I could have spent it on something else. As my long-suffering partner can attest, I'm not that into clothes. My 1950s self would have been happy clothing was cheaper, and the next time I needed to buy new clothes I'd have been pleased to spend a lot less than I used to have to for a jacket. But I'd have been likely to use the money I saved on my surprisingly cheap jacket to go out for dinner with friends or to go and watch a rugby game rather than on buying another jacket. I would have increased demand for restaurants or rugby games. Both dinner out (chefs, waiters and waitresses, and cleaning services) and going to a rugby game (ticket sellers, food and memorabilia stalls, stewards and security, stadium staff and players) require plenty of workers. Increased demand for these sectors from punters like me, with more money to spend thanks to automation making clothes or other goods cheaper, means these sectors would need to hire more chefs, waiters, stewards and security guards. Moreover, in the long run bigger stadia are built and sports leagues grow – with all the associated employment. And of course, some of those new chefs or security guards could be former textile workers. In other words, automation in one sector actually helps *create* jobs in other sectors because it increases what everyday consumers can afford to buy in the economy – even with their existing pay cheque.[28] This doesn't guarantee there are enough jobs overall, but it is a force for job creation that comes directly from automation.

The other-sector productivity effect is good news for workers. But you might think that its impact relies on everyone like me happening to prefer to go to rugby games and restaurants, which hire a lot of workers, rather than buying products from sectors that require fewer workers. What if the thing I really wanted to spend my money on was 3D-printed toys? Thankfully for workers, the nature of economics makes it, in the long run, likely that we spend a greater share of our income on sectors that are lower-productivity and employ more workers.

The cost of things like clothes, radios or televisions as a share of what we earn has fallen dramatically over time. This is because the world has learnt to make these manufactured products much more productively and cheaply.[29] Now if things like clothes and televisions from highly productive sectors keep getting cheaper, over time we end up spending a smaller share of our income in those sectors without any deliberate intention to do so.[30] That leaves us with an increasing share of our income available to spend in other sectors. Our spending is therefore likely, almost by default, to increase in less productive sectors that require, by definition, more workers.[31] This is a big reason why we should not fear widespread unemployment from technology any time soon. This pattern is also exactly what we see across many rich countries; the share of spending on the services sector (generally lower productivity) has increased a lot whilst the share of spending on the manufacturing sector (generally higher productivity) has declined.[32] Employment too has shifted from manufacturing to services.

The two types of productivity effect, home-sector and other-sector, kick in at slightly different times. In the short run, as we have seen, employment may rise in a sector with increasing automation. But this home-sector effect does not last forever. Yet in the long run the economic incentives push us towards spending more of our income in other sectors that are more labour-intensive, thereby creating jobs. This can more than compensate for the losses in employment in the automating sector. Strikingly, various studies suggest that automation actually increases employment in aggregate.[33]

New jobs thanks to the upgrading effect and new-task effect

To understand the upgrading effect, let's enlist the help of Jack Lasenby, one of New Zealand's greatest children's authors. As a kid I loved reading his books about Uncle Trev – a farmer from Waharoa with a flair for storytelling to cheer up his sick nephew. Among Uncle Trev's tall stories is a plan to pull the South Island up to the North Island and flip it over to create a nice flat farm. All this, he reckons, could be achieved with a few barges and a good strong team of horses. Uncle Trev had a knack for exaggeration, but he knew well the importance of working animals to farming in New Zealand in the past. Ploughing and harvesting were done with basic machinery and horses. It was tough work, but machines and horses made it much easier than harvesting without either. Prior to this, huge teams of people would harvest by hand. With a team of horses it only took one hard-working farmer, one Uncle Trev, to drive the animals forward with the harvester attached. The invention of tractors and better harvesters and their introduction to agriculture then sharply increased agriculture productivity. It still took one person to drive the tractor forward, yet they could get a lot more done each day. The introduction of tractors is an example of the upgrading effect. There was limited additional displacement of workers; the teams of people harvesting by hand were already long gone.[34] Yet the additional productivity of tractors made agricultural products cheaper, and made everyone who spent some money on agricultural products – that is to say, everyone – a little wealthier. This money could then be spent on other things like going to the theatre or the rugby, or taking a holiday, which was good news for workers in entertainment, sports and tourism.

As a kid in Nelson I was the (proud) owner of a Jack Russell terrier, Sam. His photo still adorns the fridge of the family home. For the last few years I've lived in Washington, DC. Staggeringly, within three blocks of my house there is not one, not two, but three 'dog grooming and training' establishments: The Life of Riley ('a luxury dog hotel'), Fur Get Me Not ('Certified Professional Dog Trainers provide both group classes and in-home private consultations'), and City Dogs ('bathing, brush, ear cleaning, nail clip, trim feet, face & sanitary and full haircut'). It's a world of doggy riches that would have been unthinkable for Sam.

Dog care and pampering is a growing industry supporting many small businesses. But it is only one example of the extraordinary array of new tasks in which humans have a comparative advantage that we have invented and commercialised. Some are now so familiar that we think nothing of them. Competing with the dog hotels in my neighbourhood are gyms, each with a small army of svelte personal trainers ready to shout at me to do one more rep. Back in the early 1960s gyms were for elite athletes and if you told your parents you wanted to work as a personal trainer they'd have had no idea what you were on about. The growth of gyms and the personal-fitness industry has been remarkable in New Zealand too. Les Mills senior started his first gym in Auckland in 1968. Having being wowed by the aerobics phenomenon in the USA, Phillip Mills introduced group classes, added music, and then added weights too. Gym instructors and personal trainers grew strongly as a profession: today there are about 5,000 fitness professionals in New Zealand.[35]

Dog groomer and personal trainer are not the only new jobs and new tasks – there are many others across a wide variety of skills, education and pay levels.[36] Consider this plethora of new job titles: search-engine optimisation specialist, data scientist, podcast producer, mobile app developer, sustainability manager, drone operator, admissions consultant (to get your kid into the top schools), user experience designers, artificial-intelligence system trainer, artificial-intelligence explainer (to tell the rest of us what is going on), employment brand manager (people want to work for ethical companies), cloud architect (cloud computing is big business now), content moderator (keeping tweets and Facebook posts appropriate), well-being coach, offshore wind-farm engineer, chief listening officer (the customer is always right), information security analyst, 3D printer technician, clean car engineer, motion graphic artist, site-reliability engineer (no one likes an error page), solar photovoltaic installer, and genetic counsellor (what do my genes mean for my kids?)[37]

These jobs aren't just invented purely by chance. New technologies also create new jobs – as the array of high-technology job titles in the above list attests. Plus there are economic incentives that push this process along. Automation itself makes inventing new tasks in which labour has a comparative advantage more profitable because there is now more labour available. New tasks, which are often generated through market-creating innovation, are a critical part of why employment stays high despite technological advances. One estimate suggests that about half of employment growth in the USA between 1980 and 2010 can be explained by the introduction and expansion of new tasks and job titles.[38]

Historical, global and psychological reassurance

Where does this leave us? This workhorse model is in fact a good representation of economic history since the industrial revolution. More recently too, various studies looking specifically at the employment effects of new automation technologies have found that they have actually increased employment in the economy as a whole or at least kept it steady.[39]

Throughout history, and today, workers have indeed been displaced by new technologies (the displacement effect). However, this has been counterbalanced in part by everyone having more purchasing power, thanks to cheaper products, and therefore buying more (the productivity effects). It has also been balanced by the ongoing upgrading of existing machines, which makes goods cheaper still and thereby ramps up our purchasing power further without displacing many workers (the upgrading effect), and by us finding new tasks and jobs to do, such as wind-farm engineers, personal trainers and my neighbourhood dog groomers (the new-task effect). In this way different industries rise and fall in their importance to the economy and yet employment on average stays high.

This is exactly what has happened in New Zealand. In 1901 in New Zealand 40 per cent of the working population worked in the primary industries: essentially farming, forestry or mining. By 2006 the primary sector employed only 7 per cent of the workforce.[40] While this was a huge change for many families, the country as a whole did not go hungry or without timber or minerals and employment remained strong.[41] Manufacturing in New Zealand had a similarly tumultuous twentieth century. At the beginning of the twentieth century about a third of the workforce was in manufacturing. That rose strongly and outstripped the primary industries by the 1970s. But by 2006 manufacturing accounted for just 12 per cent of the workforce. The service sector has grown to take up this slack. These transitions needed a lot of new policies to support and smooth the transformation. Even then, they were often painful. But today labour-force participation is at all-time highs and unemployment is low.

This brings us to an enormous problem with most studies, and much of the debate, about automation and its impacts on work. They skip over all the economic effects of automation and technological change that are good for workers. Instead, they focus on the displacement effect. That is like predicting the chances for the New Zealand cricket team at the next World Cup only using data from games they had lost. The picture would be pretty worrying, but it would also be misleading. Not only do Carl Benedikt Frey and Michael Osborne leave out the (big) issue of technological adoption, they don't include the effects that are good news for workers.[42] In New Zealand reports from CAANZ, Infometrics and PwC do acknowledge that new jobs will be created. However, the only effect they actually model is the displacement effect, predicting the percentage of jobs at high risk to be 46 per cent, 31 per cent and 24 per cent respectively within the next few decades (for these, they also imply that what can be automated will be, irrespective of cost or regulation). Media attention has, unsurprisingly, focused on these alarming numbers. But by ignoring the effects that are good news for workers we get much less insight into what is really likely to happen.

It's not just the lessons of economics and history that are missing. It's also a well-rounded version of the future. There are a number of big changes on the horizon that will profoundly shape the chances of people to find work. The McKinsey Global Institute identifies seven global trends, each of which is also present in New Zealand.

First, rising incomes, particularly in emerging economies, will create major new demand. People in developing countries want the same standard of living we have – and that means rising consumption in the developing world, and in Asia in particular. Rising individual consumption will generate about three-quarters of global consumption growth through to 2030.[43] For New Zealand, tapping into that will generate jobs at home. Second, despite what we might tell ourselves, we Kiwis are getting older. The median age of a New Zealander rose from thirty-six years in 2006 to thirty-eight in 2013, and has continued to rise since. As a result there will be far greater demand for health care and personal care – employment-intensive sectors. This also means that a smaller share of the population is going to be of working age and looking for a job at all – another reason why scare stories on automation can be misleading.

The technology sector also offers increasing job opportunities. New Zealand already has almost 100,000 people employed in the sector, about 5 per cent of the country's workforce.[44] If technological advances are anything like what is foretold, that will surely grow substantially. A fourth opportunity is the demand for workers to deliver major infrastructure projects like the City Rail Link and the possible second harbour crossing in Auckland. A fifth is particularly relevant in New Zealand: investment in housing. The Ministry of Business, Innovation and Employment estimates that through to 2022 alone New Zealand needs 50,000 to 60,000 additional builders and related tradies. There's a lot to do: the housing shortage, Kaikōura rebuild and earthquake strengthening in Wellington, to name a few.[45] New Zealand is also going to need to make substantial further investments in renewable energy and energy efficiency. Constructing environmentally friendly new buildings and retrofitting existing buildings will be a major source of job creation. Green construction alone could create 6.5 million jobs globally by 2030.[46] Moreover, in New Zealand 40 per cent of our total primary energy supply is renewable.[47] Not bad by international standards, but if we are going to meet our goals and commitments under the Paris agreement on climate change, major investments and changes will be required, bringing a whole set of new jobs.

Finally, assuming no drastic changes to economic policy, marketisation of currently unpaid work will create jobs. Tasks like childcare, cooking and cleaning are often unpaid and done disproportionately by women. Apps like Green Acres Go and Goodnest are helping to make it easier to hire people to support these tasks. Similarly, rising female labour-force participation in New Zealand is likely to hasten this trend.[48] Women's participation has been increasing steadily for the last twenty years but remains only 76 per cent, compared to 86 per cent in Iceland, and around 80 per cent in Sweden and Switzerland.[49] If more women are at work, demand for paid home support will increase further.

These trends, along with a full range of economic effects from automation (both bad and good for workers), should be included in any effort to understand the future of work. Doing so is very difficult. The McKinsey Global Institute has attempted this for six countries: the USA, Germany, Japan, Mexico, India and China. Their estimates, while still uncertain, are therefore more credible than most. The results are striking. For advanced economies they predict full employment in 2030 – no matter which scenario they run.[50] That is to say they find a world in which the vast majority of working-age Americans, Germans and Japanese go off to work every morning – just like they do today. To be clear, they suggest this will still involve a lot of people changing jobs (which will be very challenging) as well as changing what they actually do at work. The McKinsey authors suggest that between 3 and 14 per cent of the global workforce may need to switch occupation. People will need support as they transition. But still, full employment in 2030 is a very long way from stories of looming mass unemployment.[51]

Unfortunately, they do not extend this model to New Zealand. However, OECD studies have suggested New Zealand is one of the advanced countries least susceptible to automation due to our relatively small manufacturing sector and high share of employment in professional and managerial occupations. (Despite what some might hope, it's still very hard to automate your manager.) We are certainly less susceptible than Germany, Japan and the USA. Yet we should remember that more limited ability to automate may also mean lower productivity growth, which in turn means smaller increases in living standards.

These forward-looking models are supplemented by analyses of recent history. The Asian Development Bank found that across twelve Asian countries between 2005 and 2015, new technologies meant that those countries could have produced the same economic output with 101 million fewer jobs (displacement big time), but higher demand from wealthier people (the productivity effect) and whole new kinds of jobs instead meant that employment increased by 33 million to 134 million jobs.[52] Overall, new technologies actually increased employment. These findings should offer a dose of reassurance to New Zealanders that mass technological unemployment is highly unlikely any time soon.

To be clear, there is uncertainty about which jobs will be affected by future technological change, about how these changes will map onto existing social, economic and ethnic inequalities in New Zealand, and about the scale and speed of impact. This book remains sceptical about many of the more dramatic job-loss predictions and the prospect of widespread technological unemployment. But it is still entirely understandable to be concerned about these issues. Making the best of technological change and mitigating its more damaging impacts is not easy. Above all, it will require careful policy choices, which are discussed further in Chapter Five. Wise policy choices, after all, have been critical to the reasonable job we have done in the past to help people benefit from technological change.

Less alarmist views of the future of work (like this book's) are sometimes dismissed as examples of optimism bias– that is a deep-seated bias towards optimism affects our judgement, in this case about the dangers of automation.[53] Yet there is also psychological evidence to suggest that people have a hardwired tendency for 'magnifying risks in troubled times and rose-tinted false optimism in happier times'.[54] There is enough trouble in the world today – from American politics to Brexit – to suggest magnifying risks might currently be the greater danger.

A more relevant psychological risk may be the availability heuristic: that we estimate the probability of something happening based on how easily instances come to mind. News media is heavily biased to report scary news regardless of the topic: from robots taking your job to plane crashes. This makes it easy to summon up examples of bad things happening and thus to overestimate their likelihood. Studies indeed have shown, for example, that we vastly overestimate the prevalence and risks of crime.[55] The greater risk may be that we overestimate the dangers of the future of work and technology and forget the possible benefits.

Painful transitions and new challenges

That is not to say everything is swell and we are on track. We've been through big workforce shifts before. But dealing with these shifts has been painful. Very few people look back fondly on the difficult decline of manufacturing in the 1980s in New Zealand. Unemployment rose through the 1980s to 11.6 per cent in 1991, before declining again.[56] If there are indeed widespread shifts in employment as a result of new technologies like artificial intelligence, they will likely also be difficult for many New Zealanders. Losing a job, searching for another, and retraining to do so is a daunting prospect. That remains true even if there are new opportunities amongst the upheaval – particularly if new job opportunities are in different cities to where jobs are lost. Moreover, the difficulties of transition are unlikely to hit everyone equally. Instead, they may exacerbate existing inequalities around race, gender and class. Government policy to support people through job transitions, discussed in more depth in the next chapter, will need to pay close attention to how employment upheaval maps onto existing societal cleavages and disadvantages.

How difficult such a transition might be depends on how quickly displaced workers are re-employed. If this is sluggish, unemployment could rise in the short to medium term before the economy adjusts and unemployment falls. This in turn will depend on how well labour markets work, how easy it is to retrain and how much help workers have in finding new work. Put differently, government policy will really matter to how this transition goes, particularly if the changes hit hard.

This transition may be difficult even if we end up in a future of high and decent employment. Yet there are also worries, and some evidence, that long-standing features of our economic landscape may be changing as well. This suggests there may be problems even beyond the struggle of a major employment transition.

Three questions are particularly worrying. Are robots starting to take our pay cheques even if they aren't taking our jobs? Are most people going to be left with the low-paid jobs as the robots and the lucky few do all the lucrative tasks? And even if we still have 'a job', are robots making these jobs less secure with fewer hours than we might want?

Are robots nicking our pay cheques?

Too many New Zealanders set an alarm, wake up early, commute to their full-time job, work hard all day, and yet at the end of the week struggle to pay the bills and feed their families. Joanna, who was recently profiled on the Stuff website, works in retail in Henderson.[57] She earns $20.40 an hour, $4 above the minimum wage.[58] After tax, KiwiSaver, union fees and $20 extra savings for Christmas she takes home $550 a week. Rent alone is $530. She gets by with a $70 accommodation supplement from Work and Income – which, in a Kafkaesque twist, she has to reapply for every three months. Her two kids are at university but living at home; each pay her board of $140 per week, all they can afford as students. That, she says, puts food on the table and keeps the lights on. She can't afford the parking where she works so she bikes.

New Zealand has too many working poor like Joanna.[59] Our long-held, basic belief was that if you work hard you will do fine. It is under strain.

Wages have been pretty stagnant in New Zealand for the last ten years. Better wages would make a big difference to people like Joanna. For a start it would mean she could afford to park at work – at least on rainy days. For others it would mean the chance to rent a warmer, drier home. For parents it might mean the chance to buy better, healthier food for their family. Or it could mean Mum or Dad could make it to a dentist to try and fix that painful toothache.

Workers have been hit by a double whammy. Productivity growth, as we saw in Chapter Two, has been weak, so the economy isn't producing much more per person in the first place. That's a big constraint on New Zealanders getting better wages. But on top of that, workers have been slowly getting less of the income that is produced. Economists call this the labour share of income. So workers have been getting a declining share of a pie that's only growing very slowly. That is a raw deal.

The idea of the labour share of income is quite simple. Businesses make money. Some of that is paid to workers. That's the labour share. Another part is kept by the owners. That's the capital share. Historically the labour share stayed pretty much the same year to year. In the early 1980s in rich countries it was around 55 per cent.[60] This meant that when businesses got more productive and made more money, workers on average got richer too. They still got about 55 per cent, but of a bigger pie. But in the 1980s, things changed. The labour share of income in advanced countries started declining. It is now about 4 to 5 per cent lower than it was in the 1980s. In New Zealand the fall has been worse still: a decline of almost 10 per cent since the early 1980s.[61] A 10 per cent decline in labour share is a big loss. If labour's share had remained what it was in New Zealand in 1981, the Council of Trade Unions says Kiwi workers in 2018 would on average receive $10,800 more per annum.[62] And that's just for 2018 – every year since 1981 they would also have earned more.

The big questions are: what's going on, and what does it have to do with robots? The decline in labour share has unified economists in worry but divided them in explanation. Three main culprits get fingered in the pages of academic economics journals: technological change, globalisation and policy. Outside of academic journals, the fingering is more obvious: robots, China and Reagan/Thatcher/Ruth Richardson.

The scariest possibility is that while robots may not be taking our jobs they may be taking at least a cut of our pay cheques (and giving them to their owners). One influential version of this story is quite simple. Robots got better and cheaper, so businesses bought more of them. This increases how much companies produce, but most of the returns to this increase are used to pay for the investment in the robots themselves, meaning workers may get a smaller share. Yet for this to be the case, it is critical that it is sufficiently easy for businesses to substitute a machine for a worker.[63] Some argue this is now the case, but the evidence is hotly contested.[64] It's also not clear that businesses have actually been buying more robots and machines when these drops in labour share have occurred.[65]

Another story favoured by Autor is that middle-skilled workers (such as those in sales, office and administrative support and factory and machine operators) are losing their jobs to technology and the jobs they can pick up instead are worse-paid with a lower share of labour.[66] Yet in New Zealand, as we will see below, that simply does not fit with the evidence. The share of people in middle-skill jobs is declining, but happily it is the share of people in high-income jobs that is growing in its place.

The globalisation story is very similar, but in the typical telling, it is the Chinese rather than robots taking the middle-skill jobs. Removing thousands of well-paid jobs to China or other developing countries can hit the share of income going to labour pretty directly.[67] New Zealand though has had less offshoring of manufacturing than the USA for a simple reason: it already had far less manufacturing at the point at which globalisation really ramped up and China entered the World Trade Organization in 2001.

The third explanation starts with a question: how is the share of income for workers really decided? The Economics 101 answer to this is that competitive markets mean workers are paid according to their productivity. But there has long been another answer: that the workers' share is the outcome of negotiations between workers and owners of capital. The name of the game is power. The stronger the workers' position the more they get, the weaker their position the less they get. This was essentially the view of the big names in nineteenth-century economics: Adam Smith, David Ricardo and, of course, Karl Marx. For Peter Lyons, a New Zealand economics commentator and author, this power explanation is just as relevant today. Some of New Zealand's most profitable industries, like big rest-home operators, supermarkets, fast-food chains and parts of agriculture and horticulture, he says, pay the most miserly wages. Wages for low-income jobs, he suggests, are largely determined by the bargaining power of worker compared to employer.[68]

In New Zealand, the workers' share took a hammering in three short episodes. Two of them indeed had a clear backdrop of political turmoil. Muldoon's wage and price freeze in the early 1980s foreshadowed one precipitous drop. Then in 1991 Minister of Finance Ruth Richardson doubled down on Roger Douglas's reforms through the Employment Contracts Act, which seriously cut workers' bargaining power. Union membership almost halved, from 52 per cent of employees in 1989 to 26 per cent in 1996.[69] The workers' share of income plunged again. Labour's share of income declined again between 1999 and 2002, though this is less easily pinned on policy changes. Nevertheless, for Bill Rosenberg, the chief economist of the Council of Trade Unions, these policy changes are the key story.

Researchers elsewhere tell a similar story of declining unions and changes to labour-market institutions precipitating the decline.[70] In the USA a related story of policy failure is also offered: a rise in monopolies. Without real competitors, businesses are able to mark up their prices far more than in the past and in doing so they increase the share of income that stays with them and reduce what goes to workers. This may be a failure of regulators to ensure real competition. One possibility is that technological change may have also made it easier for firms to carve out a monopoly position due to network effects – think Facebook or Uber, where everyone being part of one network is a critical part of the business model. Regardless, they suggest, regulators have so far failed to rise to this new challenge.[71]

Much as we might prefer neat conclusions, this debate is not settled. There may be truth in each explanation of what has happened so far. Still, looking ahead, even if automation for particular jobs does reduce the labour share for those workers, as some economists suggest, that need not result in further declines overall. One thing that would help stop this from becoming an overall decline is the creation of new tasks and new jobs in which workers have an advantage. Our ability to do this may be the reason why, until recently, the labour share has not declined substantially.[72]

Blaming a declining labour share only on technology also has the (perhaps convenient) effect of reducing our sense of responsibility for it. This constrains our sense of agency and ignores the role that we have in shaping technological development. Even if automation does put pressure on the labour share, there are things we can do to put pressure back the other way. For one, policymakers can actively incentivise innovators to focus on creating new tasks and jobs for humans. Similarly, we can ensure that our policies on unions, monopolies and labour rights strengthen workers' position.

We also know that productivity growth, despite some weakening of the relationship, is still a powerful tool to help workers to earn better wages.[73] And as this chapter has shown, it is absolutely possible to have rising productivity without mass technological unemployment. We also know that it is possible to have a much higher labour share with technology similar to ours – Denmark has a labour share that is more than 10 percentage points higher than New Zealand's, and theirs is rising.[74] We can help increase what workers like Joanna take home. Sorting out the causal debate should be a complement to, not a substitute for, doing so.

Are robots taking all the good jobs?

When you hear jobs may be at risk from technology, the natural reaction can be to ask whose job, my job? Journalists have certainly been asking themselves this. And it's not just journalists that are being told to worry. The middle class is about to be 'blindsided by artificial intelligence', says Vivienne Ming of Singularity University.[75] In *The Future of the Profession: How Technology Will Transform the World of Work*, the Susskinds make a similar argument; robots are coming after middle-class professional work.[76] It is a common refrain that what is distinctive about this time is the threat to middle-class jobs.

Yet this feature isn't unique. The threat to middle-class jobs was the central story of the industrial revolution. Prior to the industrial revolution, if you wanted clothing, furniture or tools you sought out a highly skilled artisan. Skilled artisans were the archetypal middle-class workers of the late eighteenth century. As machines arrived, suddenly unskilled workers – and women – could work in mechanised production lines making all manner of things that used to be the sole premise of skilled artisans. The smashing of machines by Luddites that followed was arguably not so much a protest against machines as 'collective bargaining by riot', as the historian Eric Hobsbawm put it.[77] Artisans were angry at being undercut by unskilled workers. Of course the industrial revolution was a period of upheaval and difficulty. But it also laid the foundations for rapidly rising incomes and the more affluent lives most people live in New Zealand today.

Nevertheless, the dominant story of technological change through the 1980s and 1990s was that it reduced the need for low-skilled workers but increased the need for more skilled workers. If you had a decent qualification, you were fine. If you didn't, you should think about upskilling. Tertiary education surged in many countries.

Today that theory is changing, and it is a little more subtle. The new thinking suggests that it is routine tasks above all that are at risk from machines. Many low-wage occupations retain a lot of routine tasks and so many of these may still be at risk.[78] However, quite a lot of middle-class occupations include many routine tasks, while some lower-wage occupations actually include non-routine tasks. Many manufacturing jobs – often decent middle-class jobs – include routine work. Even lawyers and accountants do a good deal of routine work. On the other hand, lower-wage occupations like personal-care work and building maintenance are largely not routine.

When I spoke to Autor this was his big concern about the future of work. 'It's not really the number of jobs per se,' he said. 'It's the number of jobs that will offer a good standard of living, some employment security, and some notion that people are not interchangeable.'[79] Indeed, middle-skill jobs do seem to be declining. In the USA middle-skilled-occupation jobs have declined.[80] In Europe in the 1990s and 2000s at least sixteen countries have seen between 5 and 10 per cent declines in the share of people in middle-paying jobs.[81] And even in New Zealand, in a similar period the share of people in middle-wage jobs declined by 5.5 per cent.[82]

Yet, and this is where the story gets more complex but less scary, declines in middle-class jobs may not be bad news for the country as a whole. It all depends on what kind of jobs are growing in their place. In the USA, the evidence is concerning; the growth in jobs has been largely in low-wage work. Higher-paid jobs, particularly in more recent decades, have grown but only weakly. For a worker who lost their job in car manufacturing in Detroit, finding low-paid work as a security guard was plausible. Finding work as a professional, even if they had the qualifications, would be much tougher.

Europe often gets lumped in with the American analysis because it has also suffered declines in middle-wage jobs, but the picture is actually substantially better. Unlike in the USA, in most European countries the biggest growth in jobs has happened in high-income occupations.[83] Nevertheless, there has also been growth in low-paid jobs. So there is job polarisation of a sort in Europe (a decline in middle-wage jobs relative to low- and high-paid employment), but it is skewed to the end we would prefer – higher-paid jobs.

The news from New Zealand is much better. The share of people working in high-education jobs in fact rose by almost 10 per cent between 1991 and 2011, to 44.9 per cent.[84] Even more strikingly, the share of people in low-education jobs actually fell, although not as much as middle-education jobs. If you measure by salary levels, the story is the same – high-paid jobs increasing their share of employment in place of low- and medium-paid jobs. We know these people aren't leaving the labour force – labour-force participation is at an all-time high. We also know unemployment is low. So these workers must have shifted into higher-wage and higher-education jobs. The real story in New Zealand is therefore a very positive one: the economy has helped middle-wage earners, and perhaps some low-wage earners, to become high-wage earners. Not everything in New Zealand's economy is rosy, but this is excellent news.

Transitions like this are, at least in the long run, exactly what economic history suggests happens as technology improves and people get richer. This trend has continued since 2011: the biggest growth in occupations since then has been of managers – the best-paid occupational category.[85] None of this guarantees things will stay the same. Job polarisation is an important possible trend to keep an eye on in New Zealand, and some of what we see overseas is concerning. But the best evidence we currently have in New Zealand is reassuring rather than alarming.

Worries about labour's declining share and job polarisation have also raised fears that technological change is causing inequality. Clearly, the decline in labour's share of income – perhaps partially attributable to technology – is contributing to growing inequality. Those with a lot of capital (by definition, the already well-off) get a growing slice of the pie while workers get a shrinking one. It's also easy to imagine that rapid automation might cause a further decline in the labour share, especially if new tasks are not created for workers. Similarly, because technologies allow companies to access much bigger markets over the internet, it can be easier for one technology company to dominate vast markets and get very rich at the expense of others. Few people, after all, buy the second-best app, or use the second-best search engine, when the best one is just as easy to access.[86] This explanation is most persuasive for the very wealthy, like the owners of Google and Facebook. Still, there is a real risk that in these ways further technological advances could exacerbate inequality.

Yet technology does not explain everything. For example, the share of all wage income in the USA that goes to the top 1 per cent has almost doubled from 7.3 per cent in 1979 to 13.4 per cent in 2017; it's hard to see how robots explain this.[87] Technology, after all, is similar in Switzerland, Sweden and the USA, yet inequality in Sweden and Switzerland, even before tax redistribution, is substantially lower than in the USA.[88] Quite simply, there are other, more obvious, explanations for rising inequality: the rise of the financial sector and astronomical executive pay, for example. In the USA, CEOs earned 312 times more than the average worker.[89] In New Zealand, CEOs earn 30 to 50 times more than the average worker.[90] That's still a dramatic gap – and it has been growing.

Moreover, a lot of wage inequality is about differences in wages *within* the same occupation.[91] The technology on offer within one occupation doesn't differ much, but many firms may be failing to use the latest technology. Technology adoption differences, rather than technology itself, may be driving inequality.[92] Separately, a lot of inequality is a result of big differences in people's level of education – and large pay premiums for more education. Reduce educational inequality and we could reduce income inequality. Once again, while there is definitely a risk that rapid technological change hastens inequality, much depends on our policies in areas like education, unions, competition and labour rights.

Is technological change making work insecure and intermittent?

Odetta Ntezicimpa is a mum of four, runs a driving school part time, and drives for Uber in Auckland. With the competing demands and shifting schedule of her work and home, the flexibility of Uber works perfectly. Cate Bryant, a mother of two from Devonport with long experience as a corporate marketing executive, has been freelancing as an online marketing and sales consultant for years and counts The Warehouse, Vodafone, Rodd and Gunn, and State Insurance among her clients. These two are all about the positive side of independent work: the independence and flexibility.[93] Independent workers who are paid by short-term task or assignment are sometimes referred to as part of the gig economy, particularly when they use online platforms to find tasks. For many of those freelancing, the income from doing a few extra tasks is also a great boost. The gig economy also helps consumers with convenience and good prices – whether for an Uber ride or for help putting together furniture.

Yet the story is not all positive. Online gig platforms like Fiverr have been criticised for glorifying working excessive hours and, along with other platforms like TaskRabbit, for low pay rates. This is a common problem: a British study found that about 700,000 gig-economy workers were paid below the minimum hourly wage.[94] The most troubling form of the gig economy is tasks that are both found online and done remotely, like on Amazon Mechanical Turk. Where competition is global because the work is online, pay can be astonishingly low.[95] In New Zealand even the in-person gig economy can be a tough world – particularly if you are young, and relying on it alone to get by. Jay Jefferies's story, in the *Otago Daily Times*, is revealing. On and off for a total of six years he has been a 'temp worker'.[96] That's not quite as high-tech or exciting as the gig economy, but the reality is in fact similar: work is not certain every day and pay can be low. In short, the potential downsides of the gig economy are also clear: lack of benefits and security, potential underemployment, and in some cases poor pay.

All this means that much depends in New Zealand on what kind of gig economy we have and who works in it. For people with other obligations like studying, raising kids or a part-time business the flexibility to top up earnings can be fantastic. And for people with strong experience, like Bryant, there can be a sweet spot of good pay and flexibility. For businesses it can also provide valuable extra flexibility and efficiency.

But for those with little experience, having to rely on gigs or temp work full time, the value of flexibility is far lower and the likelihood of poor pay and not enough hours far higher. Moreover, despite hopes the gig economy might help close the gender pay gap, evidence from Uber suggests the gap stubbornly persists.[97]

Yet there are some reasons for optimism about the extent to which the benefits are outweighing the costs. Research in the European Union and USA suggests that about 70 per cent of independent workers do so by choice rather than necessity and about 55 per cent use it as supplemental rather than primary income. Independent workers also report greater satisfaction in many aspects of their work life than traditional workers. For many, the gig economy is working fine.[98]

Still, there are also longer-term potential downsides to the gig economy. One risk is that workers get stuck in temp or gig work and can't get anything more secure or longer-term. Elizabeth George of the University of Auckland found evidence in France of exactly this.[99] For people at the lower-skill end of the gig economy there is the risk that they learn little from gig work, making it harder for them to move into other work. Workers in the gig economy so far have very little power to demand better pay. This undermines their own pay, and the fact businesses can outsource to gig workers also undermines the bargaining power of their full-time staff.

Freelancing and temp work existed well before the phrase 'the gig economy'. Indeed, the very creation of the gig economy and temporary work is not just about technology; it has been deeply influenced by policy choices that have allowed and supported growing temp work.[100] But platforms like Fiverr and Uber have made gig work easier to find. This has raised the number of people working in the gig economy – and brought some of the problems out into the light. Yet the scale of the gig economy is almost certainly smaller than its media profile. In fact in the OECD on average the share of workers in temporary employment has actually gone *down* between 2000 and 2017.[101] In New Zealand there has been a strong increase in people who are self-employed without employees – but not all of these are necessarily gig-economy workers.[102] Above all, we need more information in New Zealand about whether people are satisfied and freely choosing the gig economy for some top-up money, or are stuck with it as the only option for their primary income.

Nevertheless, the trick will be to find a way to put a floor under the insecurity and low pay of the gig economy, without eliminating the benefits of flexibility and independence it allows for both individuals and businesses. The ball is in the politicians' court. One good step would be to help empower gig workers organise so they can demand better pay. Some have started; a number of Uber drivers in New Zealand went on strike in May 2018 in protest at their low pay.[103] But the big question is whether gig workers are really self-employed or whether they are employees of the platforms on which they sell their services (or something in between). That has major implications for their rights to a minimum wage and sick pay, to name just a few. Another approach is to try to create minimum-wage laws for gig workers directly: New York has just imposed a minimum wage for Uber drivers. Policy, again, matters.

Alongside the growth of the gig economy, there is also the risk people are working but can't get as much work as they would like, much like Jefferies. He is part of the underemployed. There were 112,300 people like him in New Zealand in 2017.[104] Like many deprivations in New Zealand, underemployment hits some groups more than others. Almost 70 per cent of the underemployed are women. They are also most likely to be Māori or Pasifika and to be young with few qualifications.

Underemployment in New Zealand is relatively high by OECD standards at 4 per cent of the population in 2017. This is a worry. But it is not getting much worse. The rate is only marginally above what it was in 2004 when it was first calculated. Nor is it clear that if it were to rise it would be a result of technological change. The gig economy, offering as it does the chance to add a few extra hours here and there, could help reduce underemployment.

The longer-term risk is that more and more of the work in the economy is broken down into tasks and outsourced to the gig economy. This would be a very different kind of economy, with far more of the work being done by people with fewer rights and benefits. If we moved towards much greater 'gig-isation' of the economy it might instead increase the risk of underemployment. A watchful eye is needed. If underemployment starts to spike, labour-force participation starts to plunge, or unemployment goes outside of what we're used to, we would be wise to look again for the fingerprints of technological change.

Don't panic, but do shape the future

There are reasons to believe that the future, even with rapid technological progress, might still see a lot of people working. Our workhorse model suggests that as technology improves, many things we buy will become cheaper, allowing us to spend an increasing share of our income on things that require human labour. Similarly, economics gives hope that we will keep inventing new tasks for humans to do. Yet there is a risk that while work may remain common we might also see growing inequality, stagnating wages for many people, and an increase in contract work with limited benefits and security.

In past eras of technological change we have made big choices, such as dramatically expanding secondary education and establishing laws to protect workers, to make sure that the future of work we move towards is a better one. Without proactive efforts to shape the future of work again, the world we drift towards may be fraught with problems like inequality, insecurity and low pay – even if there is plenty to suggest we will not run out of work altogether. As previous chapters have shown, despite some ominous headlines, panic is neither necessary nor helpful. But over a longer time frame we will likely see significant technological change. Now is the time to think about what we need to do to create the future of work we might actually want.

To that end, the next chapter will sketch out a broad future for New Zealand at least twenty-five years from now, in which there has been substantial technological progress but, despite this, decent, reasonably paid work remains a central part of most people's lives. Shaping such a future will not be easy. The chapter will underscore six big challenges that would need to be overcome for us to get there.

Yet not all work makes us happy. And it is possible that an alternative economic system and social model in which we work far less might bring New Zealanders greater life satisfaction and well-being. In Chapter Six I will suggest what this radical future might look like and the big challenges that New Zealand would need to surmount to reach it. Both these futures are broad representations. They are examples of a range of possible futures that could have more or less work. Yet contrasting these two helps highlight the real trade-offs and challenges that exist in shaping New Zealand's future in one direction or another.

For the next chapter, let us take *Financial Times* columnist Tim Harford's view as our starting point: 'I am confident that we would all benefit from an economy that creates real jobs which are sociable, engaging, and decently paid ... none of us should be giving up on making work work better.'[105] What would it take to do this in a world of greater technological change and automation?

Five
Making work work

Humans are underrated.
– Elon Musk, Twitter

Paul stumbled out of bed and groggily pushed the button to open the blinds (voice-activated gadgets, he found, were too much like having a third person in the bedroom). Natia was already up, at her (surprisingly pricey) early-morning small-group yoga class. These little lie-ins were one of the advantages of moving to Miramar, he mused, thanks to his nearby job at Weta Digital.

Samu, ten, and Olivia, four, were up too; Hannah, their house helper and tutor, had arrived and was getting them breakfast: organic Wairarapa blueberries and some of that yoghurt handmade in Karaka Bay. At first he'd thought it was a little extravagant having a full-time helper and tutor. That would have been unthinkable when he was a kid. A bit too reminiscent, he felt, of Natia's favourite TV show from when they were dating, *Downton Abbey*. But, as Natia had insisted, it was much better than having the kids sat in front of Netflix, or with their heads forever stuck inside those virtual-reality helmets playing some game. Plus Hannah's tutoring really seemed to be making a difference to how Samu was doing at school. Natia was right of course, as always. And he didn't feel too bad about the echoes of *Downton Abbey*. Hannah's salary hadn't been that high to start with, but the going rate seemed to be rising every year and, fearful of losing her, they'd agreed to regular raises.

Hannah's efforts also meant both he and Natia could more easily work full time. He had a lot to do after all, thanks to the booming demand for special effects in Bollywood, and increasingly in Pakistani and Bangladeshi films. It also helped Natia to work full time for her new 'slow medicine' cooperative. The idea had baffled him at first – a nurse or doctor that comes to your house and promises *slower* appointments than anyone else. Bizarrely, they sometimes even go when no one is sick, just to check in. (These days he couldn't help but think he felt baffled more often than he used to.) But the business was going well. Yesterday Natia had told him that five new doctors and nurses in Wellington were joining the slow-medicine cooperative. People really did seem willing to pay to feel close attention from a doctor. Personalised care grounded in the Māori idea of manaakitanga (taking generous care of others), Natia insisted, was what people really needed.

'You have a call from James, brother,' said a voice suddenly, interrupting his musings. He still didn't really like Alexa barging in like that but, having won the battle to keep it out of the bedroom, he had lost the battle to keep her (it, he corrected himself!) out of the bathroom. It always spooked him slightly that Alexa knew which room in the house he was in. Their house was dotted with electronic sensors – all wired up to each other, smoothing everyone's flow around the house. Tech was surprisingly good value these days. All their friends had done the same.

'Aren't you going to pick up Uncle Jim's call?' shouted Samu down the hall. 'Put him on hologram, I want to see him.' Paul demurred. This, he guessed, wouldn't be a happy call.

James, his older brother by seven years, had worked since the 2000s at a Finlay & Sons factory in Auckland preparing seafood – oysters, scallops, even caviar – every way imaginable. Not holding a degree, James started out on the production line but eventually they stripped his hours back as new shucking machines were introduced. Yet thanks to the newfangled breeding methods, and the hyper-efficient machines on the production line, Finlay & Sons dropped their prices for oysters. Kiwis took advantage by indulging their tastes heavily.

Paul remembered when 'Oyster Fridays' suddenly started at that great seafood spot in Oriental Bay. Paul knew the red-tide algal blooms in Nelson Bay, the Hauraki Gulf and Foveaux Strait had something to do with all this cheap seafood, but he preferred not to dwell on it. Still, the boom continued, and soon Finlay & Sons built another production line, which James helped supervise. James also topped up his income doing occasional gig work putting together people's new furniture. But last week he'd confided in Paul that his supervisory role was going to be mostly taken over by some remote sensing machine.

Stepping out of the bathroom, Paul picked up his phone. 'Any news on the third production line you mentioned they might get you to set up,' Paul said by way of greeting. 'No, Indians don't care for oysters from New Zealand apparently,' James told Paul bitterly. 'Fat lot of good that free-trade deal did,' he added. Paul didn't mention the deal he had clinched last month in Mumbai to work on a trilogy of Bollywood films loaded with Weta's finest special effects. 'I'll be out on my arse soon,' said James. Their conversation turned to the retraining schemes the new government was funding. There were also still some jobs in construction that didn't require as much retraining, Paul thought. But it might be hard work. 'Gotta run,' Paul said lamely, biting his tongue on the construction jobs and wishing he'd had something more useful to say.

'Byeee,' called Olivia as she danced out of the house. Hannah was dropping her down to Montessori. They'd paid for the extra-small playgroup where screens were banned. The term fees were so expensive that Paul was slightly obsessed to make sure she didn't miss a single session. Hannah drove her down – well, she sat behind the wheel at least to keep an eye on things – each day.

'Come on, Samu, I'll walk you to the bus.' Paul wasn't having his kid show up at school in an Uber, like some kind of little emperor. And now the council had put a driver back on the bus – after that horrible incident with the boy with epilepsy with no adult nearby – it was definitely the best option.

'Hurry up, Samu, the bus will be there in five minutes, thirty-six seconds,' he called. 'And don't forget you have your voice coach in Seatoun after school,' he added, flicking through the news on his phone. More spending on education, health, and something called 'active labour market policies' in the latest Budget, it seemed. Unemployment, the government was trumpeting, was back down to 5.3 per cent. 'Some people have a job but not enough hours though,' Paul muttered, thinking of James. Still, he thought, the fall in unemployment was a relief, especially after the scary rise that seemed to overwhelm the welfare system a few years ago. The regular strikes and protests had faded too.

Paul arrived to work a little later that morning, carrying his lovingly prepared flat white from Two Brothers Coffee. It was run by a couple of burly former miners turned baristas. Workers were already swarming all over the construction site next door to Weta Digital – a business consulting firm he recalled.

The rest of the week looked busy, Paul thought, grimacing as he scrolled through his calendar. Plenty more business development with Bollywood, but at least he had an interesting call with a Nigerian film studio. His days would be long – competing internationally made that hard to avoid. And late this evening he would be pounding the Miramar hills under his personal trainer's watchful eye. On Friday he and Natia were going to try out that new comedy club. If only he could help James into a new job, he mused, turning to his desk to tackle the reams of waiting emails.

Understanding Paul and Natia's world

Paul and Natia's world is a vision of what New Zealand *could* look like, assuming that technological progress and productivity growth picks up in at least some important sectors.[1] It is not a prediction, but an inevitably partial sketch through the eyes of one family of the kind of world we could aim for in New Zealand. It is based on the lessons of history and economics, and the assumption that government meets some of the big challenges that are discussed below.

As the preceding chapters have made clear, widespread technological change will not necessarily happen quickly. This chapter considers a world in which there has been significant technological progress that has automated some things but also created new tasks for people to do. For this reason, this should be considered a discussion of a possible New Zealand that is well into the future – at least 2045 – rather than just around the corner. It is a world of automation and technological change, but one in which government still clearly aims for a society and economy structured around work, high employment and inclusive growth. Let's take a look in a little more detail at what this world might look like and on what basis it has been imagined in this way.

Paul and Natia live in a New Zealand that has been experiencing technological progress and high productivity growth in some sectors. It is also a future New Zealand where most people are in work. They are clearly well-off in this fictional example. But they are not absurdly or exceptionally so for two university graduates. A sustained period of technological progress and productivity growth alongside high employment should make most New Zealanders substantially better off.

The jobs in this future, drawing on the economics of the last chapter, come from four places. First, they come from creative and innovative sectors that benefit from technology but are less susceptible to complete automation, such as Paul's work at Weta Digital and other creative industries like music, film and design. If technology allows them to market and produce their services more cheaply, opportunities from overseas – perhaps including Bollywood in India and Nollywood in Nigeria – are likely to grow.

Second, jobs come from high-productivity-growth industries, like Finlay & Sons seafood company with their new breeding methods and shucking machines. Job growth is thanks to technology delivering lower production costs and lower product prices. Lower prices in turn stimulate demand in New Zealand (Paul's favourite Oyster Fridays) and abroad. This demand creates more jobs despite automation thanks to the home-sector productivity effect. This pattern may help many New

Zealand industries, particularly if they can find new markets abroad. However, most of New Zealand's primary sector is likely to see fewer jobs even if automation allows cheaper production and prices, and stimulates demand. There simply isn't enough land (or resilience in the environment) to allow total production to expand enough to outweigh automation, which has and will reduce the number of people needed to produce a litre of milk or a kilogram of lamb.[2] That said, another form of agriculture could create jobs thanks to wealthier consumers' taste for unique artisan products, like Paul's family's breakfast of local blueberries and yoghurt. These kinds of products are marketed precisely for their sustainable, caring and human-led production.[3]

The third way people are likely to find work is in whole new high-tech tasks and jobs that have been invented thanks to new technologies. Government policy supporting market-creating innovation in particular would help ensure these jobs grow. In New Zealand, one example of this today is the jobs created at Dexibit, which uses big-data analytics to help improve the visitor experience at museums and exhibits.

Fourth, employment in Paul and Natia's world is strong in personalised services, like tutoring by Hannah and caring. Natia's job as a 'slow doctor' with extra time for her patients is intended to presage this trend. Care work in general is likely to increase with New Zealand's ageing population. Similarly, there may be a broader shift towards more personalised services. People are increasingly willing to pay for a human connection. In the realm of medicine, researchers suggest this could also have real health and well-being benefits.[4]

Somehow I (along with many others) was once forwarded an advertisement for an after-school tutor for Chris Martin and Gwyneth Paltrow's children. The salary was good but I wasn't even remotely qualified: they wanted Greek and Latin as well as the ability to provide sailing and tennis lessons. Theirs is an extreme example, but it illustrates a broader fact; as people get wealthier they spend more on personal services provided with a human touch.[5] Most people would like to have more personal tutoring for their children, more personalised and less rushed health care, aged care with greater human contact and time, personal fitness instructors rather than fitness videos, bespoke design (and redesign) of their houses, and attentive personal service at restaurants with creative chefs.[6]

We have already seen, for example, vastly increased interest in eating out and creative food and drink as societies have grown richer. In New Zealand we spent $11.2 billion on eating and drinking out in 2018. That is nearly double what we spent in 1997 (even controlling for inflation).[7] In 2017, seven new hospitality businesses opened each day.[8] And it takes just a short walk down memory lane to realise that the quality of food and drink in New Zealand has improved almost beyond recognition. The personal human touch matters here too. More expensive restaurants have personalised table service rather than relying on people to go and get their food from a counter. High-end restaurants don't shout out numbers.

It is often ignored, but the opportunity to interact with other people is a core part of the appeal of going for a coffee, to a restaurant, or to farmers' markets and boutique shops (which have grown strongly despite the rise of Amazon).[9] This desire for the human experience is very real. Even Starbucks instructed their baristas to focus on no more than two drinks at a time after customers complained that they had reduced the art of coffee-making to an assembly line.[10] Paul and Natia's world of full-time tutors, personal trainers, small yoga classes, small playgroups at Montessori and voice coaches may be the direction we are heading. This trend is one part of what would keep employment high in this world.

What we actually do at work is also likely to change. McKinsey Global Institute estimates that less than 5 per cent of occupations will be fully automated by 2030, but they suggest that in about 60 per cent of occupations at least one-third of the current tasks could be automated. That's a big change to our day-to-day activities. It will take some getting used to.

Social, emotional, creative and advanced cognitive abilities are all likely to be increasingly important. Similarly, the ability to apply expertise, interact with stakeholders and manage people are all going to matter even more.[11] Can you quickly build connections with people? Can you read the room and understand if someone is upset or struggling, and figure out why? Can you show empathy and pull a team together to complete a complicated task or hit a tough deadline? Can you combine ideas from different places to find a new way of doing things? Paul's success persuading Indian filmmakers to sign on with Weta Digital is one example of the value of these kinds of skills.

In part because of these changes, in Paul and Natia's world there is a clear value placed on education. The importance of education in the face of rapid technological progress – the race between education and technology – is already widely acknowledged today.[12] Education's importance is likely to grow further in a world of rapid technological progress where high employment remains a priority. Consequently, this vision of the future has the government spending heavily on education, including for retraining people who are out of work or may soon be, like Paul's brother James. Paul and Natia put a lot of emphasis on their children's education, from the private tutor Hannah to Montessori for Olivia and voice-coaching classes for Samu. Their focus on small groups and communication for their children reflects the fact that interpersonal skills and social and emotional reasoning are likely to be increasingly important to finding a job as technology takes over more routine tasks.

Aiming for a world of continued high employment despite rapid technological progress will not be easy, as James's situation attests. Paul's flick through the news also alludes to unemployment having been higher recently and to nationwide protests. History tells us that accelerated technological progress and productivity growth is likely to imply more rapid occupational churn, with people needing to look for a new job and potentially spend time out of work doing so. That is never easy.

Occupational churn may also lead to periods of underemployment for people – particularly as technology could further break work down into tasks, making it easier to hire people part time or on a task-by-task basis through the gig economy. Like James, people in jobs with a lot of routine tasks may need to retrain.

Decent wages in Paul and Natia's New Zealand

One claim about Paul and Natia's world sticks out: the idea that their helper and tutor Hannah, as well as other personal-service workers, will be reasonably well paid. This has big implications. If personal-service workers are not well paid, there is a risk that the class of working poor grows. Or, even if their wages are not poverty level, that inequality grows. Some of that will depend on policy and community choices like the minimum-wage and collective-wage agreements, which are discussed further in Chapter Seven. But what else is going on?

In general, economists find that wages grow in line with how much workers produce per hour – their productivity. Increasing productivity is still a great way to increase wages. If productivity increases rapidly in some sectors, as this vision of the future assumes, it is likely to do so in the sectors where technology could most easily make a big difference, such as digital effects, information technology, transportation and manufacturing.

Yet, just by the nature of the work, personal-service jobs like care nurses, waiters and teachers are less likely to see rapidly rising productivity. Does that mean their wages will inevitably fall further and further behind the wages of people in high-productivity industries like digital effects, information technology and high-tech manufacturing? A brief digression to the economics of string quartets may help us see why not.

William Baumol, an American economist, pointed out in the 1960s that musicians weren't getting any more productive but were getting paid a lot more. It still took the four musicians of a string quartet the same amount of time to play Mozart's Haydn quartets as it did in the 1860s. (Rushing it, music teachers will assure you, is not a good idea.)

Yet their pay in the 1960s was far higher than it had been in the 1860s. The reason for this, Baumol observed, was that if a performing-arts centre had paid its musicians in 1960 the same as they did in 1860, the musicians wouldn't stick around. They'd be chucking in their violins for jobs in better-paying sectors where wages had risen thanks to productivity growth. Performing-arts centres therefore have to raise wages (and consequently ticket prices) to keep their musicians from quitting for work elsewhere.[13]

This relationship holds not just for musicians but more broadly. Barbers, for example, complete the same number of haircuts an hour as fifty years ago but get paid far more today. In fact, this relationship has been documented in many countries and industries. In New Zealand, the Productivity Commission finds solid evidence of this effect over the last twenty years.[14]

Worldwide, in education and health, productivity growth has been low but costs have risen because wages have had to at least keep in touch with wages in other sectors.[15] Indeed, in general, as productivity rises in one sector along with wages, it helps pull up wages in less productive sectors. Pay therefore stays more closely related than we would expect between highly productive and less productive industries. If it did not, young people looking for jobs would opt overwhelmingly for higher-paid jobs until there were so few people willing to work as barbers, musicians, teachers or nurses that wages would start to rise and the gap would close anyway.[16]

Due to the cost increases it implies in education and health care, this has been dubbed Baumol's cost effect. But the flip side is a rise in pay for workers in sectors like health and education. For our purposes let's therefore call it Baumol's wage effect.[17]

Of course, the Baumol effect does not address all forms of inequality. Inequality *within* the same occupation is a major driver of rising inequality worldwide; so are declines in the labour share (the proportion of income paid to workers).[18] The Baumol effect helps reduce differences in wages across different sectors, but it does not ensure labour's overall share of income stays high relative to the share going to the owners of capital.[19]

Yet Baumol's wage effect is very important in explaining why growing differences in productivity across occupations – which are likely to occur in the future – need not lead to spiralling inequality.[20] However, it can be undermined by bad policy. For wages in productive sectors to drag up wages in less productive sectors, workers and young people considering their career options must be able to choose either sector. They can only do that if two conditions are met. First, they must have the skills and education to get a high-productivity, high-pay job. Second, there must be enough jobs on offer in high-productivity, high-pay sectors.

Paradoxically this means that, to get decent wages for barbers, teachers and carers, we need enough people with the skills to plausibly *not* become barbers and to instead have the option to join the next generation of highly productive industries – whether that be technology companies, agriculture or high-tech manufacturing. To do that, we need an outstanding education system. It also means that there needs to be enough demand for whatever high-productivity sectors produce – digital effects, electronic goods, hologram technologies – so that those sectors are regularly hiring workers. For Hannah in our fictional scenario, her decent wages are intended to reflect the Baumol wage effect alongside robust investments in education from the government and strong demand for workers in productive sectors in the economy, in part thanks to trade deals.

There is also a potentially vicious circle here. Ensuring enough young people have the education to plausibly choose a job in a high-productivity sector is critical to Baumol's wage effect. But the likelihood of children getting that education is undermined by rising inequality and falling social mobility.[21] Data from six New Zealand universities shows that 60 per cent of the students accepted into law, medicine and engineering in the past five years came from the richest third of homes, while just 6 per cent came from the poorest third.[22] If inequality rises and social mobility decreases, there may not be enough young people who can plausibly enter high-productivity sectors for the Baumol wage effect to keep working.

The absence of the Baumol wage effect would make inclusive growth much harder. It would mean that the wage gap between high-productivity-growth sectors and lower-productivity-growth sectors would increase, and inequality would increase with it. Higher inequality in turn would make it even more difficult for the next generation of poorer kids to get the kind of education that would allow them to enter higher-productivity sectors. This is the vicious circle. It could easily undermine the decent wages envisioned for personal-service workers like Hannah in the future.

There is also a risk that if our education system and welfare system are unable to equip enough young people with the necessary skills, some of these higher-productivity sectors simply will not exist in New Zealand. This might leave us more equal – but much poorer as a country.

In today's economy, it is education rather than job opportunities in higher-productivity-growth sectors that seem to be the bigger threat to the Baumol wage effect.[23] Around the world employers are struggling to find people to hire with the right skills. Almost half of global employers in 2018 reported talent shortages. New Zealand employers were no exception.[24] Many advanced industries are also struggling to find the right talent. Among the top-ten most in-demand skills in New Zealand are engineers, technicians, management executives and information technology workers.

Globally, manufacturing has traditionally been a high-productivity-growth sector and today has job openings that require far more skills than in the past. In the USA there are likely to be 3.5 million new manufacturing job openings through to 2025, but two million of them may go unfilled due to a lack of skilled workers.[25] The story is similar in New Zealand in the digital sector.[26] There were 14,000 new jobs in the tech sector in New Zealand in 2016. But there were only 5,090 graduates in the tech sector the year before. Tech is a classic example of the kind of sector that could pull up wages in lower-productivity sectors – the median salary for IT workers was $82,000, compared to $49,868 for all jobs in the economy. But that will only occur if more young graduates are qualified to get jobs in the tech sector. In short, our education system isn't keeping up. For Paul and Natia's world to succeed, it absolutely has to.

In a future with rapid technological progress, the challenge of finding people with the right skills will likely continue and the demands on the education sector will be even greater. But generating sufficient demand to consume everything that these newly productive sectors can produce could become more of an issue. If there isn't sufficient demand, Larry Summers points out, focusing only on skills may be like 'whistling past the graveyard'.[27]

Challenges on the road to Paul and Natia's world

Daron Acemoglu told me that he was a 'conditional optimist' about the future of work.[28] Assuming technology does progress rapidly, if governments made the right choices, the future could be one that is richer, in all senses of the word, for most people, and employment could stay high. Paul and Natia's world is a New Zealand version of this. It is not without complications but it may still be very attractive to many people. That's the optimism.

The condition to Acemoglu's optimism, he explained to me, is governments making the right choices, because there are some big challenges to meet. Six challenges stand out to me: innovating our way to good jobs, supporting people through job changes, delivering the education and skills needed, getting people to work, maintaining sufficient demand, and not destroying the environment while doing all of this. The remainder of this chapter will outline why these matter and how big the challenges are. As Acemoglu suggests, there are indeed things that can be done to help meet these challenges. In Chapter Seven, we will come back to what those things might be.

Innovating our way to good jobs

The first challenge has been well documented in Chapters One and Two, but it is worth underlining: to get to a world like Paul and Natia's, we need to see technological progress and innovation that creates new jobs, in addition to innovations that automate jobs. That is to say, we need market-creating innovation as well as efficiency innovation. A lot of the concern about the threat of reduced or bad work opportunities in the future stems from the possibility that technology will primarily eliminate jobs rather than also creating them.[29]

We may have some help in the broad task of new job creation. The incentives in the economic system to generate new tasks and new jobs, as Chapter Four outlined, pull in the right direction. But there is also much we can and should do to influence the kind of innovation we get. We must pay attention to the balance between different types of innovation and to how New Zealand's policies shape the kind of technological progress we get – and its impact on jobs. Efficiency innovation has its benefits but job creation is not typically one of them. If innovation skews too far towards efficiency innovation, Paul and Natia's world may prove much harder to attain.

Supporting those who need it

'The Robots Are Coming, and Sweden Is Fine.' So ran the headline of a feature article in the *New York Times*.[30] Swedish union leaders, businessmen and politicians alike seem relaxed about the prospect of further automation. The average Swede certainly seems to agree: 80 per cent of Swedes have a positive view of robots and artificial intelligence. Even more Danish and Dutch have positive views.[31] These are the three European countries most positive about robots and artificial intelligence. They are also three countries with particularly well-developed, well-funded and well-functioning social-welfare systems – even by European standards. It is a lot easier to be positive about artificial intelligence and robots, and the prospect of changing jobs, if you believe the government has your back – if, for example, you know that the government will support you financially while you look for a new job, will provide free health care and education to you and your family, and will actively and effectively support you in searching for a new job.

Strong welfare systems give kids growing up in poorer households a fair chance at getting a well-paid job. That is critical on its own terms to anyone who values equality of opportunity. Giving poorer kids a real shot at a good job is also a crucial step towards helping all those working in lower-productivity sectors to end up with better pay thanks to the Baumol wage effect.

Sadly, New Zealand is failing to give many kids in poorer households a good start in life. Fully 23 per cent of New Zealand children live in poverty. Moreover, about 148,000 kids live in 'material hardship', meaning they lack six or more things from a list of seventeen identified as important to children's well-being: things like doctor visits when needed, a warm house and two decent pairs of shoes.[32] Growing up in poverty reduces children's life chances. It makes it much more likely kids will have to go to hospital with respiratory problems. It means they are more likely to struggle at school; 68 per cent of kids from the most disadvantaged areas achieved NCEA Level 2 compared to 93 per cent of kids from the most advantaged areas.[33] And it increases the risk of health problems like heart disease, obesity and drug and alcohol addiction later in life.

As the number of people living on the streets, in cars and in motels has risen, many from across the political spectrum have made even stronger criticism of the New Zealand welfare system. The Child Poverty Action Group argues it is increasingly frayed and no longer adequate.[34] Jonathan Boston, a Victoria University academic, finds five major problems – from stingy benefits to perverse incentives – and has called for 'repair and renewal'.[35] Benefits are not linked to average wages in the economy, unlike superannuation.[36] As a result they have fallen behind. The net benefit for a single adult on job seeker's support of $215 per week is far below the $400 that a single person on superannuation receives.[37] From a more libertarian perspective, the New Zealand Initiative, a think tank, says that we need a welfare system that works.[38]

These criticisms are of the welfare system in today's economy. Yet the challenge is likely to grow. In Paul and Natia's New Zealand, with its rapid technological progress, there will be an increase in the number of people having to look for new jobs and needing support for themselves and their families while they search. We need to develop a system, as the Council of Trade Unions has put it, that supports a 'just transition' for displaced workers. New Zealand's social-welfare system is not currently fit for such a task. Readying our social-welfare system, therefore, is the first big challenge for fulfilling the vision of the future outlined in this chapter.

Learning what you need to learn

Upper-body strength, the ability to complete physical manual tasks in cold conditions, and good eyesight. Had there been an official description of the skills needed to be a plucky gold prospector in the 1800s in New Zealand that might have been it. Today, miners operate machines that do most of the dangerous work and instead need to apply advanced skills in monitoring high-tech equipment and problem-solving. Technological change has always meant the skills needed at work change. Rapid technological change would mean that we again need to develop new skills for work – and fast. Being numerate and literate may no longer be enough; employment using just those skills is likely to decline. The need for physical and manual work is also likely to decline, although plenty will remain. Machines are a long way from being able to replace the complex physical support nurses provide to patients in hospitals and needed by the elderly.[39]

So what skills will be needed? Demand for advanced cognitive skills like logic, critical thinking, complex problem-solving and reasoning is rising. Those skills pay; people that are better at them already earn more today.[40] Similarly, strong socio-behavioural skills like empathy, leadership, working in teams, building and managing relationships and resolving conflicts are all likely to grow in importance.[41] The ability to create and innovate will also remain important. And of course more advanced information technology and digital skills are likely to be in high demand. Yet the exact kinds of digital skills are hard to predict. This hints at perhaps the most fundamental skill needed in the future: the ability to learn.

The good news is that New Zealand's education system is doing well at teaching young Kiwis some of these skills, at least relative to other countries. The Economist Intelligence Unit rated New Zealand as the top among thirty-five countries at 'educating for the future'.[42] New Zealand has a particular strength in project-based learning – where students tackle a topic of interest to them personally and study it in depth drawing on a variety of different subjects. Similarly, the OECD notes that Kiwis have strong information-processing skills and higher technological proficiency compared to other rich countries.[43] This is a good start. But there is a risk we are simply the best of a bad bunch when it comes to building the digital capability, socio-emotional skills, creativity and thirst for learning that will be needed in the future.

The bad news is that these strengths are only true for some New Zealanders. New Zealand has struggled for many years to ensure students from less affluent backgrounds and from Māori and Pasifika households aren't left behind. Progress has been limited. Students from poor families in New Zealand are over six times more likely to do badly at maths than children from well-off families.[44] In general, the gap between the pass rates of the rich and the poor are among the widest in the world.[45] In fact, a recent UNICEF study found New Zealand has the second-widest gap in reading comprehension at primary level of all rich countries.[46] Only Malta stops us from propping up the table.

Schools of course don't operate in a vacuum; kids from poor backgrounds do worse on average in part because they face challenges others don't at home. Indeed by the time poorer kids even get to primary school they may be at a disadvantage. The brain's ability to learn is highest before the age of five. The foundations for future learning and the basis of social and emotional skills are also set by a child's fifth birthday.[47] Above all, early childhood is when we learn to learn. Miss this window and catching up later is tough. Moreover, to equip the generation entering the workforce a few decades from now with the skills they need, we have to get early childhood right today.[48] Good health, nutrition, security and education in early childhood make it far easier for kids to build the skills needed as they grow. Yet, as we've seen, far too many young Kiwis are living in poverty in childhood with basic needs unmet. Paul and Natia's New Zealand could be a good place to live for many people. But if the long tail of educational underachievement persists, it is unlikely to be a good place for all.

In Paul and Natia's world, having a tertiary education is also very valuable.[49] Not all jobs in the future will require tertiary education, but if history is anything to go by, there is a good chance more jobs will require it than today.[50] Tertiary education is still the main way students can attain advanced cognitive skills like complex problem-solving, as well as the digital skills that will be most in demand. Even today, New Zealanders with a bachelor's degree or higher earn a weekly income almost three times more than those without any qualifications.[51] Similarly, shifting up the scale, those with a master's degree or PhD in New Zealand earn 54 per cent more than those who only completed secondary school.[52]

In New Zealand 44 per cent of twenty-five- to thirty-four-year-olds currently hold a tertiary degree. That is bang-on average for the OECD. But the leading countries are far ahead. In South Korea, a country adopting technology far more rapidly than New Zealand, 70 per cent of that age group have tertiary degrees. In Canada and Japan over 60 per cent do.[53] Improving this won't be easy. Good performance in secondary school strongly predicts success at tertiary level, yet New Zealand's scores at secondary level are declining.[54] The need for higher levels of education to find employment may also make it harder for people with health problems or disabilities to find work, because people with these challenges less commonly complete higher education.[55]

One big piece of the skills puzzle will grow in importance in a world of rapid technological progress and high employment: retraining adults and helping them to find new work. Adult training programmes are not easy to get right, not least because adult brains are less geared to learning. Frequency of use – masses of practice – is particularly critical. New Zealand does reasonably well at getting people back into work quickly, in large part thanks to flexible labour markets. Yet they tend to return to jobs with far lower pay and status.[56] In fact, New Zealanders who are made redundant and then find new work tend to take a wage cut that is much larger than people in the same situation in other rich countries.[57]

The big problem in the future is that if people need to gain new skills to get a new job, flexible labour markets alone are unlikely to do the trick. New Zealand's spending on active labour-market policies – which include training and job-search support – is very low by OECD standards, at just 0.33 per cent of GDP.[58] And it has been declining. New Zealand does not look well set for the challenge ahead. And it's not just the public response that matters. There is a parallel challenge for employers: to commit to training their own workers more substantially.

Get education and support for re-employment right and we have a decent shot at the kind of affluent New Zealand described in this chapter. Get it wrong and there is a much higher risk of stagnant wages in some sectors, rising inequality, and people stuck unemployed and searching for work for long periods.

Getting to work

Paul and Natia live close to Paul's work at Weta Digital in Wellington. But that hypothetical scenario is not easy to achieve at all. Weta Digital executive producer David Conley spoke publicly in 2015 about how hard it was to find staff a place to live near their offices.[59] Being able to live near work, or at least commute easily to the office, is very important to a high-productivity, high-employment future. Despite the ease of digital communication, advances in flexible work, and digital offices, cities continue to grow and flourish because they enable the in-person communication of the complex ideas and know-how on which modern production is built. Even the place of advanced technology and communication *par excellence*, Silicon Valley, is notably a *physical* location. Companies make an enormous effort to locate there despite exorbitant costs.

Cities are typically the most likely place for new tasks and market-creating innovations to be invented. Workers in highly skilled cities also tend to earn more. The evidence suggests that is not just because more highly skilled workers go to those cities, but because the cities themselves make workers in them more productive.[60] So important have cities been to our prosperity that Edward Glaeser, a Harvard economist, argues that they are our greatest invention.[61]

Yet for large numbers of people to work in high-productivity, high-pay jobs, large numbers of people need to be able to live in or access productive, highly skilled cities. That is no small task. Indeed it is a task New Zealand is failing at badly. We have the least affordable houses, measured by the relationship of average income to average house price, in the rich world according to the IMF and *The Economist*.[62]

Auckland is, of course, the worst of all. The impact of housing unaffordability is made worse by its relatively poor transport infrastructure and congestion problems. A study on commuter pain found that congestion in Auckland was worse than in Hong Kong or New York.[63] Yet Auckland is also New Zealand's most productive city – and particularly in some areas centred around the Central Business District (CBD).[64] It's particularly difficult in Auckland to get people to the places where high-productivity, high-pay jobs are. This is a big problem today – let alone in the future. Indeed, when I asked Patrick McVeigh of Auckland Tourism, Events and Economic Development what keeps him up at night he told me it was getting 'more inclusive growth ... so that we don't leave bits of the city, or people, behind'.[65]

If we are not able, through housing and transportation, to get enough people to the places where these desirable jobs exist and where new tasks are most likely to be created, we will not end up in Paul and Natia's world. Instead we will, at a large scale, systematically force workers out of the high-productivity areas of cities into lower-productivity places. That risks forcing them out of the job market, making the country poorer.

Does anyone want to buy this stuff anyway?

Getting our skills, social-welfare system and physical access to work right is critical but it won't make much difference if we can't find people to buy the things we make in New Zealand. Demand, as economists call it, really matters. Take the dairy sector. If demand for milk powder overseas started to fall away New Zealand would be in trouble. About 95 per cent of our dairy production is exported. Similarly, if New Zealand lost its appeal as a tourist destination and tourist spending declined significantly, the economy would be in trouble. Even small declines in demand overseas for New Zealand products cause us problems – just look at the worried headlines every time global milk prices fall.

Yet if technological progress makes us much more productive in some sectors, we are going to have a lot more to sell. And as we've seen, one of the keys to keeping employment high in a sector that is affected by automation is to keep demand high for its goods. That way people can stay employed, even if there is less human work needed to produce one bag of milk powder, one movie or one piece of agricultural machinery. Price reductions help improve demand, but so do new and growing markets. This is especially important for a country whose internal market is fewer than five million people.[66] Notably, to keep employment up in the agriculture and horticulture sectors, which already rely heavily on exports, growing markets are a must.

The good news is that demand is growing globally. The not-so-good news is that much of that demand is going to come from developing countries in Asia, Africa and Latin America, with which New Zealand does not have such strong trading relationships. Our trade with China is significant, but many other developing countries have not been the focus of New Zealand's exporters so far, and nor in some cases have they been the focus of our trade agreements. Only 1 per cent of our goods exports went to India in 2016 for example.[67] For people to stay in their jobs even as the number of hours of work needed to produce a product decreases, accessing consumers in these kinds of markets is imperative. In this light, the recent global turn towards protectionism provoked by President Trump is particularly worrying for New Zealand.

There is also a second version of the demand problem: low demand in New Zealand. Low demand is simply people not buying or investing in enough in total to keep employment high and the economy growing. Kiwis need to buy most of what we produce in New Zealand in order to keep most New Zealanders in jobs – and we usually do. But if a lot of New Zealand's income goes to just a few people, this can break down. This is because the rich spend less of their income than the poor. Technological change could both increase inequality and mean that we can produce a lot more (and thus need to consume a lot more) at full employment. If that were the case, maintaining sufficient demand and therefore high employment could become a real challenge. As Summers put it to me:

> If you have a lot of capacity to produce things, and all that capacity is owned by a small number of very wealthy people with a low propensity to spend, you're going to have a big challenge generating adequate demand to purchase all that the economy is capable of producing.[68]

Usually Reserve Banks can step in to lower interest rates enough to make it attractive to invest. This in turn gets the investment and spending moving again. Yet, Summers suggests, a number of changes in the economy, including rising inequality, are already making it more difficult for reserve banks to lower interest rates far enough to do that. Interest rates are already at or close to zero in many countries after all. This could mean investment stays low and there isn't enough spending in the economy. This is called 'secular stagnation' and it can get ugly quickly; a flatlining economy and rising unemployment may come with it.

New Zealand does not appear to face secular stagnation today.[69] Nor is this scenario inevitable. In fact technological progress should mean there are exciting new opportunities for investment and spending. That should reduce the problem of inadequate demand. Yet technological change has also made it possible for companies like Facebook to become hugely valuable without anywhere near as much investment in offices, factories and equipment as the corporate behemoths of the past. Moreover, the capital that companies do invest in – primarily computers and machines – has become dramatically cheaper. This means that even when businesses do invest, they invest less and spread their money around less.

The worst-case scenario is the following: technological change displaces workers and increases inequality but doesn't raise and sustain investment. There would simply not be enough demand in the economy to buy everything that could be produced at full employment with highly productive workers – even with reserve banks maintaining low interest rates. The result would be rising unemployment and stagnant wages. It's hard to know how likely this is. But if it did come about, it would make Paul and Natia's world very difficult to arrive at or sustain.

The big green elephant in the room

Getting job transitions, skills and demand right would be an impressive feat. Yet if we don't make some serious progress on greening our growth, aiming for Paul and Natia's world is going to be a pretty terrible plan. New Zealand already faces real environmental challenges: from water quality to climate change, and on to erosion and air pollution. Yet the New Zealand of Paul and Natia, in the absence of a robust policy response, is likely to exacerbate them further. It is a high-productivity future where the intention of the government is to keep people employed. That could be a tricky mix for the environment. High employment plus high productivity equals rapidly rising output. Put simply, in Paul and Natia's world we will produce and consume a tremendous amount more stuff. To produce a lot more stuff you generally need to use more inputs: more electricity, more water, more metals, more chemicals, more land, more jet fuel. All those, historically, have come with rising environmental damage.

The great hope is that increases in output can be 'decoupled' from the use of environmental resources like metals, land and water, and from pollution. The most high-profile of these hopes is the idea that economic growth can be decoupled from greenhouse gas emissions. There are two types of decoupling. If economic output keeps rising and resource use rises but by less than output, that is weak decoupling. That, as we shall see, isn't enough in the long run. Second, if economic output keeps rising but environmental resource use declines, that is strong decoupling.[70] And that's the Holy Grail.

Yet finding the Holy Grail is hard. The world's record at strong decoupling is poor. If this comes as a surprise, it is probably because progress on it is often misleadingly reported. We used to only count the carbon dioxide emitted from what we produced in New Zealand. We didn't count all the carbon dioxide emitted in producing things overseas that were imported and consumed here. But those things are produced overseas because there is a demand for them in New Zealand. The USA doesn't produce much cocaine, but everyone knows it is the main reason why Colombia produces so much. The same is true of carbon emissions. It turns out that some rich countries have reduced their carbon emissions from production in their own countries, but when you consider carbon emissions based on what they consume, many of those reductions are reversed. What really matters, therefore, are emissions based on what we consume. Looking at these, the picture is not very encouraging.

Most OECD countries have achieved only weak decoupling – and some have not even achieved that.[71] In New Zealand's case, GDP grew by 55 per cent between 1995 and 2011 but carbon-dioxide emissions also grew by 40 per cent.[72]

A future for New Zealand with much higher output (thanks to high productivity growth and high employment) and higher consumption may also be one of much greater emissions. In fact, it's very hard to escape the fact that higher living standards are strongly associated with emitting a lot more carbon dioxide.[73] Yet there is some hope from elsewhere. Denmark and Germany, most notably, both appear to have achieved strong decoupling between 1995 and 2011. Whether they can keep that going and whether it could be replicated in New Zealand, which is poorer and more heavily agricultural, remains to be seen.

Weak progress on this worldwide is partly a question of our technical ability to produce things with fewer emissions, but it is also very much a political failure. There is a lot more that would be done if policy incentives required it. Technological innovation could help much more, but its success depends on the incentives to actually adopt cleaner technologies. Without them, why bother innovating?[74] And as we saw in Chapter One, the chances of technology riding to the rescue on carbon emissions are further reduced by the fact our innovation systems, as well as business choices about buying new technologies, are geared to save on labour usage more than on carbon emissions. Yet this is a result of the policy and economic environment we have created, not the actual scarcity of labour and carbon absorption capacity.

Unfortunately, carbon emissions aren't the only resource of concern. The news is worse in other areas. For example, one analysis looked at all raw materials including biomass, construction materials, fossil fuels and metal ores. It considered all the resources used both in the production of a good and in the good itself – no matter where in the world it occurred. The authors found that there has been no strong decoupling at all – even in advanced economies. We are not, in their view, getting any more output for the amount of raw materials used.[75] That's a huge problem for New Zealand if in our vision of the future we plan to produce a whole lot more stuff.

Weak decoupling of carbon emissions from output in some countries is still progress. But there are absolute limits to how much environmental damage and extraction we can do before, bluntly, we destroy the 'safe operating space for humanity'.[76] In a hugely influential study Johan Rockström and his co-authors identify nine planetary boundaries for Earth. Stay within these boundaries and we are able to do what we wish in pursuit of our social and economic goals. Yet, they find, humanity has already transgressed three of the nine boundaries: climate change, the rate of biodiversity loss, and interference with the nitrogen cycle. Transgressing these for a sustained period, these scientists find, is dangerous and risks undermining our ability to return to safe levels.

Weak decoupling, in which environmental damage continues to increase but at a slower rate than growth, is not good enough if it still exceeds planetary boundaries. Indeed even strong decoupling might not be enough to stop us crossing planetary boundaries if it only occurs once countries are already wealthy. Kate Raworth points out that even if the United Kingdom had achieved absolute decoupling at today's level of income, 'if other countries were to follow suit – trusting that growth would eventually lead them to a similar peak and decline – it would require the resources of at least three planet Earths, pushing the global economy into extreme overshoot beyond planetary boundaries'.[77] A world of much higher output, which is inevitable if we do get much higher productivity and keep employment high, could be a very dangerous world indeed.

A six-sided challenge

Paul and Natia's future New Zealand looks like a good place to live for many people. It is wealthy, with high employment and generally decent wages even for those in lower-productivity sectors, and has an emphasis on education and the value of the human touch. Aiming for such a world of high employment in the face of rapid technological change will be tough. Moving from one job to another is never easy and often has a period of unemployment or retraining in the middle. Still, as this chapter and the last have argued, there is also an underlying economic logic to the idea of Paul and Natia's world. Yet our six big challenges stand in the way of New Zealand achieving such a future.

First, we need to support innovations that generate jobs, not just automate them. Second, we need a social-welfare system that both can deal with far greater need as people rely on it to transition between jobs and can give everyone, no matter where they were born in New Zealand, a start in life that gives them a chance at jobs in high-productivity sectors in the future. Third, we need an education system, from early childhood to adult retraining, that can deliver to everyone (not just those from the right postcodes) the skills required in the future. Fourth, we need cities and transport systems that do not exclude large portions of the population from getting to the parts of town where good jobs are. Failure to meet these three challenges would dramatically increase the human misery of Paul and Natia's world and would, in particular, undermine the prospects of decent wages for people in lower-productivity-growth sectors like teaching and nursing. Far greater inequality would be the likely result.

Fifth, New Zealand needs to maintain – both at home and abroad – enough demand to keep employment high despite the advances of automation. Fail at this and unemployment will rise. Finally, we need to strongly decouple our economic output from environmental resource use and damage. Without this, Paul and Natia's world could be a disaster for the planet.

Meeting just one of these challenges is difficult, but what is needed is to succeed at all six. Yet, as I will argue in Chapter Seven, there are things that can be done to begin to address each of them if this is the kind of future we wish to aim for. But before we look at how, we should think further about whether this really is the kind of future we want. Perhaps there is a better world possible. Perhaps we could live in material wealth without the imperative to compete, work and sweat over our job prospects. Perhaps we could pursue other things we are interested in and delighted by, irrespective of their economic value. Understanding what such a brave new world might look like is the subject of the next chapter.

Six
A brave new world

It's time to start discussing what kind of society
we should construct around a labor-light economy.
– Erik Brynjolfsson and Andrew McAfee, *Foreign Affairs*

Rachel woke from her accidental afternoon nap at the kitchen table with a start. The green light on the grocery delivery box was on; the delivery drone must have come by while she was snoozing. Grocery shopping was a thing of the past. The smart fridge had it covered, and if she wanted anything special she would just (loudly) tell the fridge to add it. Highly automated farms and factories, whirring warehouse robots and a delivery drone did the rest. The bags were left in the fingerprint-secured delivery box on the porch. Her mother's tales of working as a supermarket checkout attendant seemed quaint now.

Her daughter, Lily, bursts through the door still sweaty from playing at Victory Square. Rachel had packed her off to the park while she made a few adjustments to her and Wei's share-market investments. Her mums'-group friends laughed at her interest in the share market. 'There is no way you can out-compete the trading algorithm,' Karen had said loudly. Maybe that was true, but Rachel could have sworn she was ahead in the short run. And it was fun to try anyway. Plus she could at least change the parameters for the trading algorithm: more or less risk, ethical investments only, Kiwi and Australian or Asia too.

Rachel had initially been very cautious with the diversified portfolio of shares that the state had given her on her eighteenth birthday. That had been a heck of a contentious policy when it was introduced, she recalled. Endless study tours to the Netherlands and Finland to see how it worked hadn't stopped the opposition thundering on about 'the end of work and responsibility'. Rachel hadn't really seen it as a big deal – it was only the next logical step from those employee stock ownership plans that had grown in popularity in recent years. 'Why should it be only the people *with* a job that get a share of the returns from technology?' she had said to Wei at the time.

These days getting your 'gift shares' from the government was considered pretty normal: something of a rite of passage into adulthood. Lily would get hers in four years. And she'd start to receive a basic income as well. Nearly everyone got a big chunk of their income from their shares these days. The basic income meant even if you messed things up pretty badly – like her sister Sarah who had sunk it all into a bach in Golden Bay before she'd had time to get any returns – you'd still be just about okay. Plus most things they bought seemed to keep getting cheaper as the robots got even better.

Of course some people still worked. There were the executives; she knew a couple up on the Port Hills, although most of them were in Auckland and Wellington. They were, Rachel had to admit, quite a bit wealthier. She was sometimes jealous. But they also spent every bloody day in an office. And to get an executive job required years of exams and fierce competition. Plus most of their income was taxed heavily anyway. It seemed like a lot of work for their virtual-reality movie theatres and personal-transport drones. She and Wei had a perfectly decent life without ever working full time.

Of course, robots and artificial intelligence couldn't do everything. In some harder jobs people still worked. There were still investigative journalists, psychotherapists, and of course politicians. One cousin was a teaching facilitator, and Wei's sister worked part time in a nursing home. They enjoyed the human contact, they said. Still, most people she knew did not work full time. And Rachel felt like she had plenty of human contact anyway, what with bringing up Lily full time, her mums' group, volunteering at the centre for disabled kids in Stoke, and her amateur theatre troupe that put on classics at the ancient Theatre Royal.

Wei too enjoyed his freedom from work. His father hadn't been around a lot, but Wei was a proudly involved dad. He and Rachel shared childcare responsibilities and household chores (almost) equally. Softball, e-sports, robot-design competitions: you name it, he was there. To Rachel's surprise, in the last few years he'd taken up wood carving. It started with a couple of online courses and soon he was producing his own pieces. One was now proudly installed at The Suter, Nelson's oldest art gallery. That still left plenty of time for touch rugby on Tuesdays and Thursdays, and the freedom to go for a craft brew afterwards with the team (driverless pods had ended the dangers of drink-driving). Rachel and Wei also spent long hours plugged into their virtual-reality system building make-believe civilisations, blowing up aliens, or simply exploring the world from their couch.

Recently Wei had also been preoccupied with the immigration troubles of his cousin Xi in Beijing. Immigration New Zealand's policy was to deny both the basic income and gift shares to new arrivals in the country. In practice that meant Xi needed a job offer in New Zealand in order to move – an unlikely prospect. Wei filled his time recording increasingly frustrated opinion videos fulminating against this policy, which he would send to every news platform he could think of – to no response.

'Jeffrey is here!' shouted Lily. Shit, thought Rachel, snapping back to reality; she'd forgotten her brother was coming to stay from Christchurch. 'Welcome,' she said, warmly hugging him as he walked in. Tall and greying, Jeffrey had always stooped as if he was apologising for his height. These days though he seemed to be further bent over, as if he were carrying a sack of rice on his shoulders.

Lily loved seeing him but that wasn't why Rachel had invited him. Jeffrey had been struggling recently. Normally he didn't talk about that kind of thing, but when they spoke last week he had mumbled that he was 'not doing that great'. Rachel tried not to worry, but Jeffrey finished school when society still expected people to get a job. He lost his forestry job to a machine after eight years, and was doing fine with his shares and basic income. But they both knew that not working left him at a loss, and his marriage to Ryan had collapsed shortly after he stopped working.

She hoped Lily might even provide a bit of inspiration. School for Lily was big on finding your passion. Artificial-intelligence-powered interactive computers taught most things from reading to coding, but there was lots of spare time for children to try out more or less whatever they wanted, from rock climbing to digital effects. This kind of 'radical freedom', as the school slogan put it, was to help kids find a passion, something over which they might want to build mastery. Some schools – and parents – were still preoccupied with 'skills for employment'. But to Rachel it just didn't make that much sense now that so many professions – from accounting to dentistry – were heavily automated. Why spend your whole childhood studying to get a job where you would just be required to sit, work and follow orders most of your days – especially when you simply didn't have to?

Understanding Rachel and Wei's world

Viewed from 2019, Rachel and Wei's hypothetical world can seem pretty outlandish. Yet the advances in technology are not what seem most unusual. The oddest idea is that people will simply not work. We will come to the question of just how outlandish that may or may not be. But for now let's simply observe that a surprising number of people are talking about a future like this some time quite soon. Andrew McAfee and Erik Brynjolfsson of MIT, authors of *The Second Machine Age,* have suggested it is important now to be thinking about what a labour-light economy might mean for society. McAfee believes this will happen in his lifetime. That's pretty soon: he is in his early fifties and, despite his techno-optimism, is not one of the people who believes in immortality. Martin Ford, author of *Rise of the Robots*, similarly believes we are heading rapidly toward a future with a lot of unemployment.[1]

How to respond to these possible changes is also much discussed. Richard Freeman, an eminent Harvard economist, has suggested that for workers to keep up a decent standard of living they need to own some of the robots that replace them.[2] Noah Smith, an academic and influential American economics blogger for Bloomberg, has raised the idea of receiving a diversified portfolio of equity on your eighteenth birthday. And, as we've seen, a host of technology leaders from Mark Zuckerberg to Elon Musk have talked about the need for a universal basic income (UBI) in the face of coming automation. Meanwhile Finland, Kenya and Canada have all experimented with some form of UBI.

Let's take a closer look at the ideas that underpin Rachel and Wei's world.

What is the point of having a job? The question is perhaps so obviously answered that it is rarely asked. 'To earn money, you idiot' might be a common response. But from the perspective of the economy as a whole, jobs do other things. First, they mean things get produced. Businesses need to hire workers into jobs so they can produce milk, furniture, laptops, accountancy advice and holidays – in fact, literally everything. All that we consume is produced with some human labour. Second, and less often recognised, jobs are the single most important way of redistributing money in the economy. Businesses make money and then distribute a lot of it through paying wages and salaries. This means people can afford to buy the milk, laptops and everything else we produce. Third, jobs also have a social and psychological function; they often provide an important sense of purpose, identity and companionship for people.

The distinctive feature of a labour-light world, at least as it is written here, is that it might be possible to produce many goods and services without much human labour. Yet eliminating the productive function of jobs also eliminates the redistributive function. This is the central problem of a labour-light world: the productive role of jobs has been largely replaced by machines but the redistributive role has not. If people do not get an income some other way, most will not be able to buy what is produced in the economy. Economists call this insufficient demand. When this happens the economy is in danger of shuddering to a halt.

There are other implications too. With robots producing most things, the owners of businesses (rather than workers) are likely to pocket most of what businesses make. In the absence of a strong policy response, the share of income that goes to capital is therefore likely to rise steeply. People who have invested in businesses that use robots will get much richer. And, of course, the already wealthy will have the most money invested in businesses. Inequality is therefore likely to increase.

Any vision of the good society in which most people don't work needs to find a way to redistribute income effectively. The sketch of Rachel and Wei's life near Victory Square in Nelson gives two possible ways to do this: a state-distributed set of shares or capital given to every person at age eighteen and a basic income from the state. Others favour the economist Milton Friedman's idea of a negative income tax, in which an income threshold is set, for example at $15,000. Those earning exactly $15,000 would pay no taxes and receive no benefits. Those earning above it would pay taxes on the share of their income above $15,000, and those below would pay no taxes and would receive a subsidy of, for example, 50 per cent of the gap between their actual income and $15,000.[3] Proponents argue this would provide an income floor as well as radically simplifying the welfare system.

Introducing a substantial UBI would give people the ability to choose how much they worked without the pressure of putting bread on the table. There is no suggestion that in a future like Rachel and Wei's people would be stopped from finding work if they wanted to. Yet a decision to set up a large UBI and taxes on income to support it, as well as a host of other policies and societal norms, would deeply shape individual decisions about whether and how much to work.[4] That's why a democratic debate about what kind of future we want is so important – the decisions we take collectively about policy will heavily influence our individual choices about work.

Only once significant redistribution is in place and societal norms begin to change might people choose to work less. Yet if that happens it's quite possible they would. One reason is that jobs may have begun to fail at their psychological function: to provide meaning to people. There is already a growing chorus claiming exactly this. David Graeber, an anthropologist at the London School of Economics, asserts that as much as 40 per cent of the workers in rich countries may be stuck in what he calls 'bullshit jobs'. These he defines as:

> a form of paid employment that is so completely pointless, unnecessary, or pernicious that even the employee cannot justify its existence even though, as part of the conditions of employment, the employee feels obliged to pretend that this is not the case.[5]

Graeber cites an array of crowdsourced colourful examples to support his theory. There is the receptionist who answers one phone call a day, fills a dish of mints, winds a grandfather clock once a week and ... that's it. Or the call-centre employees who pitch pricey credit reports to people who could get them free online.

A poll of British workers found that 37 per cent of workers did not think their jobs made a meaningful contribution to the world. In the Netherlands 40 per cent of people thought their jobs had no reason to exist. Yet meaningful contribution to the world isn't easy to define. I know plenty of people who work in the humanitarian and development sector who would hesitate to state a yes to the question of meaningful contribution to the world. Others have observed the same about doctors.[6] It's also hard to understand how so many bullshit jobs could still exist, given capitalism's otherwise relentless obsession with efficiency. People in these jobs may still serve a purpose to their employers, but the job may seem to lack meaning in a global frame. That's still worrying. It's not just Graeber that is concerned. Zeynep Ton founded the Good Jobs Institute at MIT precisely to address the problem of low-paid, uninspiring jobs in the USA. Bullshit jobs would largely be a thing of the past in Rachel and Wei's world. Yet, as we will discuss below, not working can also leave people feeling they lack purpose and meaning.

Rachel and Wei's world also has a very different conception of education. This may be less new than it looks. The purpose of education has not always been to get a job. In most medieval universities, education was overseen by the church, and the teachers were bishops, priests and monks. Language and arts were the subjects of the day, rather than sciences. Philosophy grew in prominence as Aristotle's work was rediscovered. The earnings potential of studying grammar, astronomy and rhetoric likely did not register highly to the average student in Salamanca, Bologna or Oxford in the 1300s. Knowledge for its own sake was paramount. Of course outside of formal education many people still studied in some way. Craftsmen would take their sons on as apprentices to train them in the family secrets. But it wasn't until much later in history that the vast majority of people received any formal education. Work, in any case, did not require much education.

Modern education is a different beast. Tertiary education in particular is framed as an investment. Employability and earnings potential cast a long shadow. Students and parents scrutinise average pay rates for recent graduates when choosing a subject and university. Governments push schools and tertiary institutions alike to produce students that align closely with the needs of the labour market. STEM (science, technology, engineering and mathematics) is the watchword of education in New Zealand, and it begins at primary school. Moreover, a whole new industry has sprung up around technical and vocational education – a sector literally named for its applicability to the job market. Meanwhile, university students scramble feverishly for internships between their years of study to increase the chances of finding paid work upon graduation.

Fears about automation have taken this focus on employability to new levels. Amid worries robots may be quite good at science themselves the new watchword is STEAM (science, technology, engineering, *arts* and mathematics). The arts are included because they might help with employability, not for their intrinsic value. Education in this conception continues to narrow towards providing only the things that are useful to employers. What gets left out are other things that might help individuals live a fuller life.[7] This looks set to continue in the short run. But in a labour-light economy like Rachel and Wei's employability may well be irrelevant for most people. Education's value in and of itself would likely come out of the heavy shade of its value to employers.

Education could again become about self-creation; it could help free us to craft who we are or might want to be. That would be a return to a more ancient approach. Schools could become cultural centres rather than job training grounds. Indeed, the word school comes from *skholē*, the Greek word for 'leisure'.[8] Yet as Rachel and Wei's world hints, this could still have value to the rest of society through the stability and happiness it could generate. The challenges for such an education system will be very different. One question might be: how do you help grow the social skills and psychological abilities to cope with, and thrive in, a world with little work? Another might be: how do you best help people to find passions that can provide them with meaning in the absence of work? A third could be: how do you support a society to retain cohesiveness and discipline when the structure and social order of work is removed?

Education is not the only area where the end of the goal of high employment would strike at the bedrock of policy. The goal of high employment currently shapes policy well beyond the Treasury. New Zealand's jobs are concentrated in places like central Auckland, Christchurch and Wellington. Making it possible for people to live near enough to their job is a big focus of urban policy. Similarly, transport planning is dominated by the need to address peak-hour congestion as people commute to and from their work. Yet if most people didn't work, neither of these would matter nearly as much. Both policy areas might change to focus more on helping people to live in and access places that enhanced their well-being. Instead of checking how long the commute to work is when looking at a house, we might focus much more on how close we are to friends and family and check the commute to beaches or parks.

In health care the pressure to get people fit and healthy to return to work quickly, and to get them off of the disability benefit, would dissipate.[9] A more empathic and considered approach to dealing with injuries, sickness and mental-health issues might follow. On the other hand there may be increased pressure on mental-health services to deal with issues arising from the absence of work. As the case of Rachel's brother Jeffrey implies, these are likely to be most acute as society makes the transition to a labour-light world. Social welfare too would be radically different. Welfare in New Zealand is heavily oriented around getting people back into work. Sanctions can be imposed, including cutting benefits altogether, for failing to prepare for work or participate in work. The end of full employment as a policy goal and the introduction of a basic income or gift shares would completely upend these areas of policy. It might also dramatically change gender roles, with men potentially taking on a far larger role in the care of children and elderly relatives than they have in the past.

It is hard to overstate the shift required to remove high employment as an overriding policy goal, even assuming automation would maintain a decent standard of living for all. And while there is growing interest in the idea, there is one oft-cited reason why it might not work: greed. A common view is that people find new ways to work and make money because they want bigger houses, new entertainment, nicer food, and to keep up with what the Joneses over the fence are doing.

We do certainly seem to keep working. John Maynard Keynes, writing in 1930 about the *Economic Possibilities for Our Grandchildren*, thought that by 2030 each of us would only need to work fifteen hours a week to maintain our standard of living.[10] Today the average full-time worker in New Zealand works 37.6 hours a week in their main job. This has barely decreased: in 1986 it was 37.8.[11] Average hours per year has declined very gradually since 1986, in part as we are able to take more holidays, but our productivity has increased far more.[12] That means we have kept working long hours and earning more, rather than taking more time off and only earning as much as we did back in the 1980s. Desire to consume more is certainly one explanation for this.

Yet this need not necessarily make Rachel and Wei's world impossible. What if we could keep consuming more and more but also stop working full time? Or, put more bluntly, what if greed need no longer be an impediment to taking time off? McAfee and Brynjolfsson suggest that technology can sever the link between our desires and full employment. In their view technology could soon do so many of the tasks in an economy that even without most people working we could see constantly *rising* economic output (not just maintenance of our existing living standards). As Chapter Seven will discuss, rising consumption may even be possible in a labour-light world with more modest technological advances, but it would be from a poorer starting point.

Yet even if technology did enable this, the problem is that for us and the Joneses to engage in an accelerated game of my-virtual-reality-cinema-is-better-than-yours, each of us needs to have enough money to do it. Robots may be able to produce televisions without any humans, but without enough humans loaded up with the cash to buy those televisions there will be absolutely no point in robots producing them all.

There is no escaping the fact that in an automated economy, for people to keep consuming more and more, most people must somehow have growing incomes. And with businesses getting the lion's share of income and most people not working, redistribution alongside continued technical progress is the main way to do that. Get that right (no mean feat) and it isn't impossible that we and the Joneses could be hurtling along keeping up with each other's growing consumption – whilst not having to work much at all. That might help people remain satisfied with Rachel and Wei's world. It could square the circle of economic growth and growing consumption alongside low employment.

Growth heretics

Yet (sacrilegious question alert) why do we actually care so much about economic growth? The obvious answer is: to make us richer. That means we can consume even more stuff, which according to the greed argument is what we really want. But in most countries there is a more immediate need for economic growth: to keep people employed.

Being unemployed in today's New Zealand is horrible. Most unemployed people are poor, and if they stay unemployed long their likelihood of earning well again plunges. Too often they also get slated in the media for being lazy. On top of this they sometimes struggle with the loss of social connection that unemployment can bring. As Andrew Oswald of Warwick University put it, 'There is overwhelming statistical evidence that involuntary unemployment produces extreme unhappiness.'[13] If this wasn't enough to make governments care, the public do care – whether out of solidarity with those already suffering unemployment or fear of becoming unemployed themselves. Rising unemployment is a death knell for governments.

Yet technology that makes us more productive allows us to produce the same amount of stuff with fewer people. This means that as long as there is productivity growth, we need to increase how much we produce – grow the economy – to avoid unemployment rising. Quite simply, if we had increasing productivity but only produced the same amount of goods and services, people would gradually be laid off. An alternative answer to why we care so much about economic growth is therefore: because we want people to be employed.

Rachel and Wei's world turns this all on its head. Thanks to the gift shares and basic income people do not need to be employed to be economically well-off. Most people, the scenario suggests, are also able to find meaning and purpose in a labour-light economy. These ideas are contentious. But taking them as a given (for now) means we no longer need to keep the economy growing to keep employment high. If we don't need economic growth to keep employment up, do we need it at all?

The importance of ever-increasing consumption on its own is much less obvious.[14] A slew of sociologists and anthropologists have pointed out that this preoccupation with novelty and material things may just be a psychological crutch (often failing) in our search for meaning in life.[15] Despite this we (literally) buy into it strongly. Nevertheless, we know that being richer doesn't quite make us happier in the way we might expect. For example, lottery winners are no happier than those who don't win the lottery.[16] Growing income does seem to improve our emotional well-being and our life satisfaction but not without caveats. For example, emotional well-being does not seem to improve any further in the USA once you earn a household income of about US$75,000.[17] 'Life satisfaction', however, seems to keep increasing with income.[18] But the richer you are, the more money it takes to get the same increase in life satisfaction. Moreover, when people are asked to predict their life satisfaction if they had a higher income they tend to overestimate by a lot compared to how satisfied people with higher incomes actually are.[19]

Perhaps most tellingly of all, one of the most effective ways to raise life satisfaction is to spend money in ways that buy you more time rather than more material things.[20] Similarly, the people who place the most importance on money and material possessions have lower emotional well-being and lower life satisfaction, while those focused on less materialistic values like community connection and self-acceptance are happier.[21] Trying hard to keep up with the Joneses, it turns out, might be making us unhappy.[22] Perhaps greed need not be such an impediment to a world where many people work far less.[23] After all, a society like Rachel and Wei's – which provides financial security at a sufficient level, places less emphasis on income as a measure of a person's value and allows a lot of free time – appears to have some of the key building-blocks of happiness.

There is one other big reason why we might be interested in a world without economic growth: the environment. Despite many decades of trying we are not much good at strongly decoupling environmental impact from economic output. In fact, as the last chapter showed, some studies suggest we haven't made any progress at all. As we grow and consume more, we damage the environment more. Yet there really are hard limits to what we can do to our planet before we properly screw it up. And we are drawing rapidly closer to them.

Decoupling is orthodox economists' great hope. But others like Tim Jackson, author of *Prosperity Without Growth,* and Kate Raworth, author of *Doughnut Economics,* do not believe decoupling is going to ride the rescue. Decoupling isn't a one-time thing, Raworth points out. If GDP keeps growing, then the rate of decoupling would have to more than keep pace with it every year.[24] That's an even tougher ask. For Jackson in particular, it just ain't happening; if we want to keep Earth within a safe space for humanity, we need to figure out prosperity without growth.

A world of high automation and redistribution where *not* working is a widely accepted norm removes employment as the motivation for growth. That opens the door a crack to a world without growth. It's up to us whether we open the door further, or pull it shut once more.

Of course, even without the job-creation rationale for growing consumption and economic growth, it may remain a priority for some. The owners of businesses and robots will make more money the more is consumed so, assuming they see that as the route to happiness, they may try to drive up consumption and growth.[25]

The other difficulty with ceasing to focus on growth is deciding at what level of income to do so. That's not an easy political choice to make. As Robert and Edward Skidelsky put it in their book of the same title: *How Much Is Enough?*[26]

The remainder of this chapter does not assume that in such a world growth will end, but we will come back to the question of growth, technological possibilities and trade-offs in Chapter Seven.

Challenges on the road to Rachel and Wei's world

A New Zealand in which most of us do not work would be a truly radical change. It would be more radical still if it heralded the end of economic growth as a national priority. Yet a future of high automation, major redistribution and low employment could be an appealing future. People would be freed from the need to work to have a decent income and from the social stigma of unemployment. In some cases they might also be freed from their 'bullshit jobs'. They would have the time to pursue their passions, connect with their children or simply indulge in their hobbies. Many thinkers and writers – and even more workers stuck in a never-ending meeting or in traffic on the way home from work – have hoped for a future in which we would not need to work. For some, there is further appeal in the opportunity to end the model of ever-increasing consumption – a possibility although not a necessity of Rachel and Wei's world – and in doing so take pressure off our natural environment.

Yet the challenges to making such a world work are very large. Three stand out: finding meaning without work, redistributing the pie and technology. If we continue to aim for consumption growth, environmental decoupling will also be a major challenge. All the environmental difficulties raised in the last chapter would also apply, although for the sake of brevity they are not discussed again here. The remainder of this chapter aims to get to grips with the size of these challenges, while Chapter Seven will suggest what we might be able to do to overcome them.

Finding meaning and happiness in a world without work

The spectre of a jobless society stalks our imaginations (and our media) increasingly brazenly these days. Yet some places have already experienced it – and forewarn of the problems we may face. In the small town of Whakatū in Hawke's Bay, joblessness arrived in October 1986.[27] That was when the Whakatū freezing works closed suddenly, putting 2,160 people out of work overnight. The lost income meant financial strife for many. This was eventually assuaged a little by the unemployment benefit and redundancy payments, which even enabled some to buy a house. Yet the damage of unemployment went far beyond the lost pay cheque. The pain was deep. In the eight years following the closure of the freezing works, a team of researchers from Otago University found that there was double the risk of serious self-harm for those made redundant compared with workers at a nearby freezing works that did not close.[28] Not everyone was affected equally; older Māori men were particularly vulnerable to depression. Some of those laid off also reached for cigarettes and booze in much greater quantities than before.

The laid-off Whakatū workers also keenly felt a loss of the whanaungatanga (kinship) that they had enjoyed at the freezing works. Work sports days, work picnics and hāngī, and after-work drinks with mates; all gone. Never again, they recounted, would they lead up to Christmas by singing carols in full voice with hundreds of their workmates as they worked the production line. The darkness of unemployment seeps into relationships too. Men in particular struggled to move on, their identity tangled up tightly in work. In some cases, divorce was the result. The closure gutted Whakatū town itself. Ten years later even the school bus had stopped.

Sadly the story of Whakatū is all too representative of the destructive and disillusioning impact of joblessness. Loss of status, loss of social connections and loss of purpose are a powerful and perilous combination. Loss of income matters greatly too – but even in countries with generous unemployment support the damage of unemployment is severe. It is genuinely difficult to overstate just how severe. A study in Germany – a country of generous unemployment benefits – found the pain of unemployment doesn't ever fully go away.[29] People never recovered the level of life satisfaction they had before unemployment – even after being re-employed. Prolonged unemployment actually can be harder to recover from emotionally than a spouse dying or divorce.[30] Partners of those laid off also find themselves less satisfied with life, not so much for the income as for the emotional stress of it all.

Unemployment in today's society is undeniably horrible. And there is good evidence that underemployment – working less than we would like – is also pretty debilitating. A survey of the underemployed in the USA found that 'those working part time but wanting to work full time are more similar to those who are unemployed than to those who are fully employed'.[31] It is not just about the money. As the late Calestous Juma of the Harvard Kennedy School put it when talking about the hopes for a universal basic income, 'Humans don't exist to shop. They aspire to have purpose in life, to enhance their competence or mastery and express their individuality through autonomy and creativity.'[32]

What does that look like – when are we at our happiest and best? Mihaly Csikszentmihalyi, a psychologist at the University of Chicago, has a clear answer: when we are in a state of 'flow'. Flow is a mental state in which we are totally absorbed in doing something and find ourselves energised and enjoying it.[33] Most people intuitively know what flow is. We just call it 'being in the zone'. People in flow can be so absorbed that they lose track of time. Yet you cannot laze your way into a state of flow. People tend to find flow when they are doing something challenging that demands the best of their highest skills and abilities.

In the words of Csikszentmihalyi, when people are in flow, 'Alienation gives way to involvement, enjoyment replaces boredom, helplessness turns into a feeling of control, and psychic energy works to reinforce the sense of self, instead of being lost in the service of external goals.'[34] That is certainly something we might all want to aspire to. Indeed, Mihaly Csikszentmihalyi and co-author Judith LeFevre find that being in a state of flow is more important to determining happiness and satisfaction than whether we are taking leisure time or working.[35]

The trouble is, for the idea of a world without work, they also find that people are far more often in flow at work than in leisure time.[36] Despite what Graeber suggests about bullshit jobs, when surveyed, people are also more positive and less anxious when at work. Yet, perhaps consistent with Graeber, we don't always recognise this ourselves and often believe we would be happier outside of work.[37] Reported positivity and happiness while at work fits with the fact that people who are employed are much happier than those who are unemployed.

Moreover, a study that included New Zealand workers found that the happiest of all are those who are self-employed: perhaps combining the psychological benefits of work with autonomy and freedom.[38]

Where does this leave the labour-light world described at the beginning of this chapter? The first fundamental question is whether we can overcome the malaise that hits us when we do not work – or even when we work less. A basic income attending to our core financial needs would surely help. But the loss of status, purpose and social interactions can also be rough. Those in favour of a post-work future argue that these are just a product of our work-fixated culture. Yet changing this culture may be very difficult, particularly if some people still work some of the time – as is likely to be the case even in a world as advanced as Rachel and Wei's.

One way around some of these damaging problems is to try and find flow more often outside of work. This is the second big psychological challenge for a labour-light world. It is a bigger task than it may seem. People who don't work today largely do not spend their days doing challenging, highly skilled things. Instead, they watch a lot of television. Surveys of time use suggest that people without work, and in particular men, spend much of their time on 'mass media activities'. That means watching the box. In 2010 in New Zealand men who were not in work spent close to five hours per day on mass-media activities. Retirees are similar. Men over the age of sixty-five spent more than five hours on this a day and, of these, those living alone spent six hours.[39] Consistent with this, retirement has been found to reduce instances of flow.[40]

This brings us to the third and perhaps most difficult question: how do we navigate the transition from a world where work is an imperative and people on benefits are often vilified to a world where most people might happily not work? That change will not happen overnight. Instead there will be a long period in which fewer people work but many others still do. For the first people who stop working (or work far less) and receive a state income, this is likely to be extremely difficult. The mentality and identity shift required for everyone to let go of work when they have been brought up with it at the core of their identity will be wrenchingly difficult – as the imagined case of Jeffrey suggests.

For Rachel and Wei's world to be a happy and satisfying place for most people, these three questions need positive answers.

Sharing the pie

Successful large-scale redistribution sits at the heart of this vision of the future. Without it this world will not exist. And if redistribution is too modest people will not have the money to buy what robots produce. That would mean a serious risk that there is not enough spending in the economy to keep production up or growing. If that happens, tax revenues would fall and the amount redistributed would also fall, making the original problem worse.

The task of building such a system is colossal. Some suggest it is impossible. Walter Scheidel, a Stanford historian, contends in his book *The Great Leveler: Violence and the History of Inequality from the Stone Age to the Twenty-First Century* that in the course of human history we have literally never reduced inequality substantially without one of four bloody occurrences: mass-mobilisation warfare, transformative revolution, state failure and system collapse, and severe disease pandemics. These four horsemen of the apocalypse, as he brands them, are in his reading of history the only things that have caused major decreases in inequality.[41]

Not everyone agrees, and there is evidence of at least relatively meaningful and peaceful declines in income inequality in the past in Bolivia, Brazil, Malaysia and Mauritius as well as in Sweden.[42] Yet there is no escaping the fact that reducing inequality substantially, for which redistribution is a primary tool, is very difficult. The reasons are myriad. One, as Thomas Piketty has argued, is that the economics tend to run the other way. He shows empirically that an old saying is in fact true; it is easier to make money if you've got money.[43] With the economics sloshing money to the rich like this, even quite vigorous scooping of money in the other direction through redistribution may not be enough to keep inequality in check and demand up.

The rich also often have the ear of politicians. Whether through campaign contributions, old-boy school connections, or familiarity with the same circles of society, the effects are similar. Taking from the rich and giving to the poor is hard. Taking from your old friend or campaign donor (against their wishes) and giving it to strangers is a lot harder. Of course, in some countries, the rich simply are the politicians. The power of the rich is no small problem for a future of large-scale redistribution. As the author Nicholas Carr puts it, 'It strains credulity to imagine today's tech moguls with their libertarian leanings and impatience with government agreeing to the kind of vast wealth redistribution scheme that would be necessary to fund the self-actualising leisure time pursuits of the jobless multitudes.'[44]

There is also the little-discussed secret of redistribution that gets in the way; people don't like it much. In New Zealand between 1990 and 2005 support for the statement 'the government should redistribute income and wealth from the rich to ordinary people' dropped from 48 per cent to 26 per cent.[45] Robin Hood has never been less popular.

Confronted with a much more unequal future, perhaps people may support, or even demand, greater redistribution. Yet there are also reasons to doubt this. At the same time as inequality increased in New Zealand through the late 1980s and early 1990s, the share of people perceiving the gap between the rich and poor to be too great actually declined (from 73 per cent in 1992 to 63 per cent in 2009). Increasing inequality, it turns out, does not necessarily mean increasing discomfort with inequality. To be clear, most people still believe the gap is too large. But this period of rising inequality was also when support for government redistribution fell to just 26 per cent. This relationship seems to hold elsewhere too. A study of eighteen advanced countries found that greater inequality in pre-tax incomes is associated with less support for welfare spending and, indeed, less actual welfare spending.[46] In New Zealand on average we've instead come to believe more strongly in a narrative of personal responsibility: that the poor are poor because they are lazy. These attitudes are a product of a lot of things, including the dominant economic narratives of the era – and could change. Growing support for large-scale redistribution in a widely automated future is not impossible. It is ultimately our collective choice. But the shift needed in attitudes for New Zealand to make such a choice is very large.

Is technology creating options or choosing our future for us?

Automation and impressive technological progress lie at the heart of this vision of the future. Many have suggested that we will have no choice but to embrace a future like Rachel and Wei's simply because technology will take all the jobs. As I will argue in the next chapter, we could aim for Rachel and Wei's world and hasten technology toward it by focusing in particular on efficiency innovation over market-creating innovation. But might we really be forced into a world without work *whether we want to be or not*? We know that in previous technological revolutions we have not run out of work. What would need to be the case for this time to be different, irrespective of what we want?

For this era of technological change to be different, three conditions would need to hold. In short, technological change would need to be (much) broader, smarter and faster. 'Broader' we might call the robots-do-almost-everything condition. 'Smarter' we might call the not-everyone-is-smart-enough-to-be-a-data-scientist condition. 'Faster' we might call the we-cannot-relearn-fast-enough condition.

Vast breadth alone could be enough to force us into a labour-light world if it were extreme enough. But more likely we would need some combination of all three conditions to truly eliminate the possibility of most people working. For example: automation is broad, so robots do most jobs; those jobs that are left are very difficult, so most people could not do them even if there were demand for it; and they require people to change jobs very quickly, so most people simply cannot keep up with the changing skill demands to stay regularly employed.

Whether these conditions are likely to be met in the next twenty-five years is, of course, difficult to judge. We should be cautious about what we know. But for guidance we can look, as this book has done, to history, to today's nascent technologies, to how technology gets adopted, and to how the economy adjusts to technological change. It is not the intention of this book to adjudicate this debate to a final answer. Instead, in Chapter Seven I will argue we in fact have more choice than we often believe about which future to head towards. Technological change does not eliminate political choice. But it is important to lay out these conditions, to have a way of thinking about the changes happening around us and what they represent. So, what does a review of these conditions suggest today?

Artificial intelligence is a general-purpose technology. This means that its uses are not specific to one industry; instead it could be applied across many. For example, it could simultaneously enable digital call-centre assistants to replace people and help optimise manufacturing processes using data from sensors on the production line. Siemens, the largest manufacturing company in Europe, is already using artificial intelligence to do this kind of optimisation in its factories.[47] Meanwhile Marks and Spencer, a British department store, is moving a hundred switchboard staff to other roles because chatbots are taking over their tasks.[48] Speech voice-recognition software from Twilio, a Californian company, and Google artificial intelligence will listen to customer requests and route their queries to the appropriate department or shop. The routing of calls was previously done by humans.

The claim made by people like Ford, author of *Rise of the Robots*, is that soon almost no task or job will be untouched by the metallic arm of a robot or the unflappable voice of artificial intelligence. This would certainly see a spike in unemployment as people look for new jobs and retrain. Yet the breadth of applications would have to be truly exceptional for it to lead to a substantial *permanent* increase in unemployment. Artificial intelligence would have to be so broad that it would nix the productivity effect discussed in Chapter Four, by eliminating every last hour of human labour needed in most jobs. It would have to be even broader and deeper to mean that AI would be better and cheaper, even for new tasks we could come up with for humans.[49] This is particularly unlikely given that the future will likely involve more personalised human services – in demand precisely because they are human. People would have to actually prefer having a robot to nurse them, or provide psychotherapy or teach their children. That's, at the very least, hard to imagine. So does the breadth of applications look sufficiently exceptional?

I do not think so. McKinsey Global Institute, for example, writes that 'little is new about the breadth of impact of automation technologies'.[50] General-purpose technologies are not new. I remember my mother using her old, battered typewriter. I also remember when she bought our first computer – a Hewlett Packard from Noel Leeming in Nelson. It was an amazing change. That remarkable transformation to using computers has arrived in a vast array of industries from finance and professional services to retail and health care. Computers are now an integral part of these industries. These computers automated whole tasks and dramatically reduced the human labour to do many other tasks.[51] This happened simultaneously across many industries. Yet employment remains high today and computers actually created millions of jobs.[52] Go further back and electricity completely transformed almost every industry: from heating to refrigeration to manufacturing. Further back still and the steam engine dramatically disrupted all manner of production from weaving to coal mining, automating away thousands of jobs that were needed to generate power. Artificial intelligence may indeed be general purpose and highly disruptive but there isn't yet much evidence that it is likely to be exceptional any time soon.

But what about the not-everyone-is-smart-enough-to-be-a-data-scientist condition? Among others, Ryan Avent, author of *The Wealth of Humans,* suggests that artificial intelligence need not automate absolutely everything to lead us to a world where most of us do not work. Instead, he suggests that if technology automated all of the jobs that are relatively easy for humans and left only jobs that can be done by very smart people, we could still end up with most people not working. Even if there were demand for more people to do very challenging work, the problem, in this view, is that most simply aren't up to it.[53] In this way, neither the productivity effect nor the new-task effect would help most people as they simply would not be capable of doing the jobs available.

In the short run this would indeed be worrisome – it is difficult to switch from being a manual labourer to a data scientist. But for this to lead to permanent long-run technological unemployment it requires the assumption that most people – no matter what support and education they get – just cannot learn the creative, entrepreneurial, social, emotional or high-level technical skills needed for the occupations that remain, even if they started trying to from childhood. The unstated implication of this argument is that many people will soon run into some sort of innate, essentially genetic, limit to what they can learn and do.

Yet it surely stretches credibility beyond breaking point to argue that any country's education system, welfare system or early-childhood support is so perfected that students' innate ability is the real constraint to greater success. Does anyone believe that New Zealand's education system is as good as it can ever be, or that it couldn't be adjusted, extended and improved if the skills required of graduates were different? The idea that the next generation might not be able to develop new technological skills is also debunked on a daily basis by teenagers: today's fifteen-year-olds are technologically savvy in a way that was unimaginable to the generation before.

Still, the flattening off of tertiary graduation rates in some rich countries is cited as evidence for the limits to what people can achieve. This is nonsense. A simple international comparison gives lie to it; 70 per cent of Korea's young people have a tertiary degree and 60 per cent of Canada's, but only 44 per cent of New Zealand's. Unless we think New Zealand's education system at all ages is basically perfect, flatlining tertiary graduation rates tell us little about the limits of young Kiwis' innate ability to learn what is needed for the future. Our graduation rates are instead a function of the job market in New Zealand, which does not yet demand advanced degrees in such high quantities (we maintain low unemployment with our current graduation rates), as well as the problems of our education system and the cost of tertiary study. None of these are unchangeable. It may not be easy, but if the South Koreans can build a system to support such high rates of tertiary graduation we could too – if we needed and wanted to. The jobs of the future may not necessarily require an advanced degree – social and emotional skills are likely to be extremely important. But even if these jobs did require advanced degrees of us all, it is profoundly mistaken to believe we humans are running up against our innate ability to reach the required cognitive heights.

What if the skills workers need simply change so fast that we cannot relearn fast enough? Computer programmers programme in a language. But talk to a programmer today and come back in a year and there is a good chance the programming language they use has changed.[54] The languages in demand change rapidly. Apple used the language Objective C when they introduced their first iPhone in 2008. This meant a lot of programmers shifted from Java to Objective C. Recently Apple's preferred language has shifted to Swift, and programmers are having to upskill to stay in demand. Keeping up with the new skills is tough for programmers, but so far most manage.

Yet if an even more accelerated version of these programmers' troubles were to become the norm in the rest of the economy too, it could be very difficult to maintain high employment. By the time people have learnt a new skill it may be obsolete. A dispiriting cycle of training and retraining with very little actual work in between could become the norm. In such a scenario, dropping out of the workforce would surely be common. The productivity effect and new-task effect would both be undermined by such churn – if you simply don't have the time to learn the new skills you need for the jobs and tasks available in a new economy, high demand for those jobs isn't much use.

Rapidly changing knowledge and skills also mean standards are not established for a given technical skill. The QWERTY keyboard only became standard in about 1900. Before that there were a huge range of layouts. When a standard approach is agreed on, schools can start to teach it and workers can point to an agreed standard and say, 'I can do exactly that.' Without this kind of standardisation, workers can't easily go job-hopping in search of better wages because competitors use a slightly different system and would have to retrain them.[55] These barriers to moving mean there is less competition for workers in the sector as a whole. In other words, rapid changes to which skills are needed can keep wages low. If this keeps happening, and low wages continue, people may even leave the sector and labour market.

The threat to employment from rapidly changing skill demand appears more of a risk than the other two conditions for inevitable technological unemployment. Yet it is much more likely to make job transitions harder, leading to particularly severe short-run unemployment, than it is to result in permanent technological unemployment. For that to be the case, the pace of change in technology and the new skills required would have to never slow down. That seems unlikely. Innovation has typically happened in bursts. At some point people get a chance to catch up. Countries with nimble and flexible education and retraining systems – and with businesses that support learning by doing and rapidly agree on common standards for new skills – will be able to catch up most quickly.

Moreover, it is unlikely that all the skills required in a technologically advanced world will be as niche as a programming language. Instead, general socio-emotional skills and higher cognitive abilities like logic, which can be used in many jobs, will remain important. Transferrable skills like these reduce the need to constantly retrain for a new role. Above all, the biggest reason to doubt this particular threat to a future with high employment is that despite individual examples like programming, there isn't much evidence that technological innovation and adoption are increasing exponentially – or even at a faster rate than in the past. As we saw in Chapter Two, adoption in particular is still a slow and uneven process, even in the internet age.

Greater breadth of automation, increased cognitive demands on workers, and faster technological change would all be disruptive. The impact on temporary unemployment could be serious, even with small shifts in these directions. However, much more dramatic shifts would be required to generate *inevitable* long-term technological unemployment. It seems likely we will see increasing numbers of remarkable applications of artificial intelligence and robots to a range of industries. But it is not clear that we are seeing anything like the extent of the shift in these three conditions that would be required for long-term technological unemployment to become inevitable. Each of these conditions is important to keep track of. Things could change, especially if our policies tried to hasten automation more quickly. But for now the prospect of a labour-light world being forced upon us by the advance of technology looks distant at best.

Variations of Rachel and Wei's world

In Rachel and Wei's world, a lot of people do not work at all, while a few executives work much more to keep the machines ticking over, and another handful work part time. Yet some people may resent the inequality of responsibility and the economic inequality this could allow, even if tax and redistribution were significant. This could make for a restless society. People may look to work more and more to try to keep up with the consumption of the executives. However, as Chapter Seven discusses, if technological progress continued and redistribution was effective, it is possible almost everyone could still experience at least slowly growing consumption. This would go a long way to supporting stability.

Yet Rachel and Wei's world is just one way we could organise society to allow people to work far less. A variation would be that most people have a job but everyone works far less often. For example, people might work only two days per week, or four months per year.[56] Another would be to aim for much earlier retirement – at say forty rather than sixty-five.[57] There is a spectrum of possibilities for how much we might work and when we might do it. Rachel and Wei's world is towards one end of this spectrum while Paul and Natia's is towards the other.

Aiming for a variation of Rachel and Wei's world changes the picture – but not that much. If we aimed to have most people work but just two days a week, many of the implications and challenges would remain. Some would be a little less severe, but others more complicated. For example, we would still need redistribution, but it would be more modest in scale. That might be easier. Part-time work might also help people find more purpose in life. On the other hand, the education system would need to focus on both employability and supporting passions and activities for all the time people do not work. That might be harder.

Moreover, job sharing would present its own difficulties. In routine jobs it may be relatively doable, but those are the jobs most likely to disappear. In professional jobs, changing the person who works every day – or even every four months – would create real problems. For a start, employers would need to find multiple people (rather than just one) with the necessary skills that fit with their team. And then there are likely to be problems and coordination troubles every time someone hands over. This is in part why today's economies are characterised by a lot of people being overworked and another set being underworked.[58] Nevertheless, alternative versions of a labour-light world like this could be preferable as futures in their own right, or could be important initial steps towards a world more like Rachel and Wei's.

Some people may view Rachel and Wei's world as a much more attractive destination than a world of continued work. But to achieve any future like theirs will require radical change. Above all, it requires good answers to the questions of how to create and sustain a large-scale redistribution system and how to find meaning outside of work. Technological progress in the next twenty-five years is unlikely to make such a world – and answering these questions – unavoidable. Much depends on our own policy choices. What is more likely is that technological change accelerates but not to an unprecedented speed. This would continue to open up the possibility of a world with less work by reducing the number of work hours needed for a prosperous society. Technology is not destiny. Instead, it increases our options while presenting new challenges. And in the last two hundred years humanity has experienced extraordinary technological change. This already gives us room to shape our future. Indeed, as the next chapter will argue, a future in which many people work far less is not unimaginable in the long run – even without unprecedented technological progress. The choice of whether we aim for a world of less work with redistribution, or instead head towards a future of continued high employment, is in our hands. The key, as the next chapter will discuss, is to understand that we can shape the future of work, and to be clear about the very real trade-offs that would be needed to achieve something akin to Rachel and Wei's world.

Seven
It's up to us

It always seems impossible until it is done.
– Attributed to Nelson Mandela

Paul and Natia's world has much in common with our world today; society is organised around people getting up in the morning and going to work. There is a commuter rush. Money comes mostly from work. People take holidays from their job. Rachel and Wei's world is very different. Income comes from the state in the form of stocks and a basic income. Finding your passion is the real challenge. Education doesn't prepare people for work but instead aims to help them find hobbies and purpose.

Considering these two alternatives, many will have one big question: which future is most likely? That is a reasonable question, and we may go toward something different altogether. Yet it is not, I suspect, the most important question. Instead, the big question is: what future do we want?

That is not a question we seem to often ask today. Yet it is the kind of question that has been asked and answered countless times in history around the world.

In the 1950s, South Korea was widely considered a basket case in terms of economic development. A USAID report called it a 'bottomless pit'. The average Korean in 1961 still earned less than half as much per year (US$82) as the average Ghanaian (US$179). Fish was one of Korea's biggest exports.[1] This year, while writing this book, I've been frequently distracted by a Korean-made Samsung smartphone. Even when I've avoided that, I have often instead flopped down in front of our Samsung television. South Korea made the policy and business choices that helped it go in just forty years from basket case to one of the richest countries in the world, known for its high technology.

In 1945 Europe was in tatters, decimated by war. As early as 1957 the Common Market was established, which eliminated customs duties, dramatically simplifying trade. By 1993, any European could travel freely across borders that were once front lines in the bloodiest war in history. In fact, the four freedoms first laid down in 1957 meant that goods, services, people and money could all move freely across Europe. That had been utterly unimaginable in 1945. Since 1972 a total of forty-eight referendums have happened in Europe to help decide on the future of the European Union. It has been created through the choices of its citizens and leaders. The Brexit referendum is only the latest of these – and, no matter what you think of it, is yet another example of how the future is not predetermined, but shaped by humans.

In New Zealand prior to 19 September 1893, women could not vote. That day, after a relentless campaign by suffragettes and a series of massive petitions, New Zealand became the first country in the world to give women the vote. New Zealand looked very different back then. It had no serious welfare support for the poor, sick, needy or old. If you hoped for anything more than a primary education, you had to stump up; secondary school was not free and almost no one attended. Retirement was little more than a dream for most. If you were poor and fell ill, you were in trouble; there was little free health care. Yet all this changed through a series of democratic choices, culminating in the Social Security Act 1938, about what life should be like in New Zealand. Free secondary education was introduced, universal health care was established, the unemployment benefit was introduced and a universal old-age pension created.

South Korea, Europe and New Zealand all made choices to shape their futures in particular ways. These transformations happened despite their respective undertakings being so enormously complex that most people considered them impossible or even unthinkable.

This chapter is about our ability to once again shape the future we want. It aims to remind us of the agency and influence we have, especially in the long term, and lays out the actions and policies that could shape our future towards either of the big alternatives outlined in this book. Yet in setting out these policy choices, it will also highlight what could go wrong, and what failure in the pursuit of either of these futures might look like. In doing so, it aims to help us better answer the big question of what future we prefer.

Jennifer Gidley, president of the World Future Studies Federation, reminds us that 'the single, predictable, fixed future that the trend modelling proposes does not actually exist. Instead what is out there is a *multitude of possible futures*' (my emphasis).[2] Moreover, the future is not predetermined by great forces, but is instead something we choose. Gidley writes that 'we all have the capacity to create our desired futures, far more than most of us realise. As the only space where we have a degree of freedom, the future is a site of great power.'[3] Indeed, for Tim O'Reilly, the Silicon Valley publisher who writes on the future of work, 'it's up to us'.[4]

The existence of human agency and choice about the future does not mean there are no constraints. There are many: social, economic and political. Some of these have become more acute in an age of globalisation. It also does not mean that success in creating a particular future we want is guaranteed. It does mean, however, that what happens in the future of work is not predetermined by some economic law or technological inevitability. Social and economic conditions can influence our actions, but the real question is what future we want to create. Given the examples of history, it is surprising how often in the debate around the future of work that this is forgotten.

This book has shown the extraordinary variety of ways that our choices, politically, commercially and individually, can shape the future of work. Without recapping them all here, we know that we can strongly influence how technology will affect work because we already see different choices leading to different outcomes. The development and deployment of automation technologies is far from uniform; productivity growth is vastly different even among rich countries, with New Zealand bringing up the rear. The deployment of new technologies in general depends much on the policy environment – it is no coincidence that Germany and the Nordic countries have far cleaner sources of energy than the USA. Sweden has had a carbon tax since 1991. Germany has a national policy for renewable energies.

Even when robots are deployed, we see very different outcomes. Increased use of robots and automation in the USA has been linked to declining labour-force participation. Yet in parts of Asia and Europe industrial robot use is even higher and yet labour-force participation is at all-time highs. Industrial robot use has been associated with rising wages in Europe rather than job losses.[5] More effective welfare states, more widely available high-quality education and retraining systems, and greater equality of opportunity in general have much to do with it. Different policy and business choices shape different outcomes.

Nevertheless, policy choices are not like flipping a switch and the light coming on. They can strongly shape outcomes, but in most cases there remains some uncertainty about the exact effects. Changing policies can also be extremely difficult politically. Moreover, the impact of policy takes time and in the short run its effect may be limited. Yet as Larry Summers put it to me, in the long run policy is hugely influential.[6] It sits behind, for example, the enormous differences between South Korea and Ghana. The public policy decisions we make today and in the near future will matter tremendously to the kind of long-term future we create.[7] Those decisions should at the very least be informed by a clear debate about what kind of future we want.

Aiming for Paul and Natia's world

To aim for a world like Paul and Natia's is to seek a continuation of modern economic history: technological advancement that changes work, but does not lead to widespread unemployment. If that's what we wanted, what is it likely to take?

It won't be easy. Six big policy areas will be fundamental if we want to end up in Paul and Natia's world: innovating for jobs, improving welfare, getting our education system to create a love of learning and to retrain people effectively, enabling people to physically access good jobs, supporting high demand, and decoupling our economic growth from environmental damage.

Innovating our way to good jobs

Not all innovation is the same. Innovation can eliminate jobs, but can also create them. As this book has emphasised, market-creating innovation can help generate new jobs, while efficiency innovation, though excellent at helping us do more with less, tends to result in fewer jobs. As Chapters One and Two showed, there is much we can do to influence what kind of innovation we get. The trick for New Zealand will be to get the right balance. Historically, the economy has balanced relatively well between the two. However, if we push incentives too far towards efficiency innovation, or see it happening regardless, it will be a lot harder to create Paul and Natia's world.

Many policy choices to affect this were detailed in Chapters One and Two; let me just highlight a few here. First, state funding for innovation should pay more attention to the kind of innovation it is supporting – are we funding enough projects that could help generate new jobs and help people work with technology rather than be replaced by it?[8] Second, taxes should not make labour excessively expensive, so that it pushes companies to cut workers; nor should subsidies make it artificially profitable to replace workers with machines. Third, we should properly price other things companies use – like carbon emissions – to help our innovators to focus on creating things that can address our real challenges – like climate change – rather than chasing labour-saving automations to excess. Finally, there is a big role for business to shift to a more long-term mindset, perhaps inspired by the Māori values of whanaungatanga (kinship) and kaitiakitanga (guardianship). This would recognise the value of workers, and businesses' responsibility to the wider community today and in the future. A long-term mindset from businesses would encourage more focus on market-creating innovations to open up whole new business possibilities, rather than just efficiency innovations to do what they already do more cheaply.

Supporting those who need it

The second big policy priority is to ensure our welfare system and safety nets give everyone a decent start in life – no matter their location, class or ethnic background. As Chapter Five explained, New Zealanders who grow up in poverty with inadequate access to basic necessities and health care have much-reduced chances of entering the higher-income, higher-productivity jobs of the future. Fail to give people a decent start and we will leave more people competing for the same jobs in lower-skilled sectors. This in turn means lower wages for those jobs. The reality today in New Zealand is that not everyone gets a decent start in life. Disadvantage often maps onto ethnicity as well, with Māori and Pasifika communities more likely to be poorer.

We already see how this can damage the economy. Researchers at the OECD have suggested, not without controversy, that New Zealand's increase in inequality since the 1980s knocked the country's economic growth back by a third.[9] This, they argue, is because too many of the poorest 40 per cent of New Zealanders miss out on the start in life, high-quality education and health care they need to have a good chance at high-income work. This is terrible for those who miss out on opportunities in life, but it is also a loss for New Zealand as a whole.

Part of the problem is that we just don't spend that much on supporting families in need. New Zealand's support spending on the unemployed, disadvantaged and elderly, at 18.9 per cent of GDP, is below the OECD average.[10] Spending isn't everything, but at some point you get what you pay for. It is very difficult to provide a strong social safety net if you don't stump up the money. Researchers at the IMF agree with the OECD's view: boosting spending on combatting inequality helps spur growth in general.[11] Spending on redistribution through services like education and health is also more popular among New Zealanders, in part because these are universal services that everyone can access.

In Paul and Natia's world of technological change, more people are likely to need income support as they transition from one job to another. Yet New Zealand's social-welfare system has gaps even at today's level of job churn and unemployment.[12] The OECD recommends that New Zealand broaden access to benefits for the unemployed, introduce longer minimum-notice periods to help give workers more time to find a new job before they must leave, and introduce mandatory redundancy insurance. Today workers can be laid off with little notice and no compensation.[13] There is no mandatory legal requirement for either in New Zealand. Here today, gone tomorrow. This is not good enough right now, and it certainly won't be if people have to change job more frequently due to technological change.

One of the difficulties for parents trying to give their kids the best start in life is balancing looking after them with a career. And it is still so often women who end up taking responsibility for this care work for both kids and the elderly. More generous parental leave could make a real difference to kids getting the best start in life, and parents, and women in particular, being able to maintain a career. This can also be a way to rebalance care responsibilities. On a visit to Stockholm I couldn't believe the number of dads out with babies on their own. It turns out there is a reason for that. Swedish couples get 480 days' paid parental leave but, critically, 60 of these must be taken by the man or they are lost. Swedish men now take 25 per cent of all paid parental leave in the country; in New Zealand the figure is closer to 0.5 per cent.[14]

Learning what you need to learn

Education is the third big policy area we need to get right to reach Paul and Natia's world. As Chapter Five outlined, New Zealand currently ranks highly on 'educating for the future' for teens and young adults.

The trouble is that not all kids get or are able to make the most of this education. Today, and for some time now, the big challenge in New Zealand is ensuring that all students, no matter whether they are from Remuera, Wainuiomata or Reefton, get that good 'education for the future'.

One of the big reasons for educational inequality is what happens outside the classroom. As Max Rashbrooke points out, school matters less than we often think to educational outcomes.[15] Kids spend roughly 950 hours a year in school, out of 5,840 hours awake. Some estimates suggest that school accounts for only 20 to 40 per cent of a student's grades. The rest is down to things like the state of their home life, poverty or wealth, parental attitudes and their early childhood experience.

New Zealand's inequalities in income, housing, and early childhood experiences are a big part of why educational outcomes are unequal. Moreover, just like these broader inequalities, educational inequalities also map onto ethnicity. Māori and Pasifika are overrepresented at the bottom. Once again we need a more robust welfare state to ensure kids do not grow up in poverty, can access the health care they need, and live in warm, dry houses. Without the basic building blocks at home, even an education system strong at educating for the future will continue to leave too many children behind.

That's not to say there is nothing that could or should improve within New Zealand's education system. The socio-economic status of students has a very large impact on school performance in New Zealand. Yet there are numerous countries – including the United Kingdom and the USA – that are more economically unequal than New Zealand but where socio-economic background has less impact on school performance.[16] In other words, our education system may do less to overcome economic inequality than many others. We can also still get better at educating for the future. At primary and secondary school that means instilling a lifelong love of learning so that retraining is less of a burden in life; it means building the socio-emotional skills and complex problem-solving skills that are less likely to be touched by automation; and it requires providing a strong basis of digital skills that young Kiwis can use to enter emerging technological fields.

Despite the claims of online-learning evangelists, teachers remain absolutely crucial to how children learn. Even with further integration of digital learning in classrooms, this is highly unlikely to change in the future.[17] According to the Economist Intelligence Unit's report on educating for the future, 'effective teachers are even more central to a successful future-skills education system than in more traditional schooling environments'.[18]

I was fortunate to have a string of excellent teachers through primary, intermediate and secondary school in Nelson. They had a huge impact on my life. Yet in New Zealand it is hard to avoid concluding that we simply undervalue teachers. In 1979 a teacher at the top of the pay scale was paid about the same as a backbench MP. Today, our backbenchers earn more than twice as much.[19] The same 'educating for the future' index that lauded our overall system noted one big weakness: New Zealand's teachers had dropped to nineteenth place among thirty-five countries surveyed in terms of pay purchasing power. It is no coincidence that in 2018 New Zealand was facing a major teacher shortage.

This is not only about what teachers are paid. It is also about how we as a society value teachers. In Finland, teaching is considered a prestigious job that comes with real responsibility and autonomy to shape how and what is taught to the next generation. Standardised testing is far less of a preoccupation for both students and teachers alike. As a result, teaching is difficult to enter; it requires a five-year master's-degree programme, which is massively oversubscribed. If New Zealand wants to continue to succeed in educating its kids for the future we need great teachers. Right now we are struggling just to get enough teachers. Changing societal attitudes takes time, but better pay, gradually increasing standards for teacher hiring, and more autonomy for teachers would be a good place to start.

We still don't know enough about what technological change will mean for the education sector. Yet throughout history there has been a strong trend towards people requiring more and more education to gain employment. As farm employment plummeted in the American Midwest in the 1900s due to automation, the farm states made education compulsory through to the grizzled old age of sixteen. That was hugely expensive – and hugely wise. It gave America an impressively productive workforce for the industrial era.[20]

This trend of needing further education for new kinds of jobs may well continue. That would mean more New Zealanders attaining tertiary degrees. Making tertiary education cheaper helps – especially for students from poorer backgrounds.[21] Improving the quality and standards of those degrees matters too – combatting grade inflation (giving students higher grades than their work would have received in the past) would be one place to start.[22] While it is hard to know the specific technical skills most in demand, it is a good bet that tertiary education will need to do better at building teamwork skills, socio-emotional reasoning and an ability to learn and re-learn. Above all, the tertiary education system will need to be agile and responsive to new technological developments and changes in the economy.

In a world of rapid technological progress like Paul and Natia's, we would also expect people to have to change jobs more often. Many of these people will need support to retrain and find a new job.[23] Yet New Zealand spends relatively little on active labour-market policies, as these are called, and focuses a large share of its spending on case management and job-search assistance.[24] Despite this only a minority of workers contact Work and Income New Zealand (WINZ) when they lose their jobs. That, the OECD observes, is because WINZ focuses largely on disadvantaged jobseekers, meaning that 'displaced workers are, to a large extent, left by their own to find a new job'.[25] About a third of our spending on active labour-market policies goes on training, but when the total is so small this doesn't amount to enough.[26] Many adults in New Zealand do participate in further learning, thanks to open access to university after the age of twenty. This is a good start, but these opportunities are not well targeted at the unemployed, who need further study or training most but are least able to afford it.[27]

Denmark, on the other hand, invests heavily in active labour-market policies. Under its 'flexicurity' model – designed to deal with the pressures of globalisation – workers get generous benefits and lots of support retraining, alongside highly flexible labour markets.[28] Danes who don't find work within a given period must participate in private-sector or public-sector work-experience programmes, classroom training or a range of other programmes, such as job counselling. Work experience in the private sector is partly subsidised by the state, and classroom training is structured into short modules, highly targeted to the job market. Denmark goes to extraordinary lengths to ensure the training is up-to-date and relevant to employers.

New Zealand has the flexible labour markets, but falls far short on generous benefits and support for retraining. Our one-legged stool doesn't serve us quite as well as Denmark's three-legged version. For Denmark, flexicurity has been a real success: low inequality, low unemployment and high-income jobs alongside relatively high job turnover. High job turnover may be harder to avoid amidst rapid technological change in the future. If so, New Zealand needs to get serious in particular about supporting retraining and work-experience programmes for the unemployed.[29]

Getting to work

As Lant Pritchett from the University of Oxford likes to say, productivity is more about place than about person. If you move to a more productive place, you too generally become more productive. This holds whether moving in the same country or moving overseas.[30]

In New Zealand, some regions such as Auckland are more productive than others.[31] Even within Auckland some areas are more productive than others. This often translates into wealth and well-being too. The Auckland Prosperity Index, which looks at broader household prosperity on a scale from 1 to 10, finds vast gaps: Ōtara–Papatoetoe scores 0.7, compared to Ōrākei at 9.8.[32] The big problem is enabling people to physically access opportunities in more productive parts of the city and country. That challenge is likely to deepen in the future.

Housing, unsurprisingly, is a big part of the problem. As ANZ Bank's research team put it, 'High house prices make it difficult for both workers and businesses to move to locations that will prove more productive, meaning that resources are not allocated in the most efficient way possible.'[33] We've got all kinds of round pegs in square holes; New Zealand's mismatches between people's qualifications and their jobs are the worst of any rich country.[34] This hurts people's wages and lowers New Zealand's productivity.[35] Because of our mess of a housing market, we aren't easily able to up sticks if we find a job better suited to us somewhere else. In a world of more frequent job change due to technology, an inflexible and costly housing market could see people stop working altogether if they cannot afford to go where the jobs are.[36]

If we can't live near our work, can we at least get there somehow? Not that easily. New Zealand's public-transport systems are weak overall, while traffic is a major problem even by international standards, most famously in Auckland. If you cannot afford a house near productive parts of a city, getting there can be prohibitively slow and expensive.

For most people to have a chance at participating in the high-skilled and high-productivity parts of the economy in the future, they have to be able to physically access them. This means improving housing affordability. Politicians seem to have been discussing forever what needs to happen to increase this. Too often, though, these discussions ignore the most basic of economics; for houses to be more affordable we either have to be much richer, or houses must have lower prices. The former will take a long time to make houses realistically affordable – some suggest half a century – and the latter will hurt existing homeowners.[37] Having let housing affordability in Auckland become the worst in the world, there are no easy choices.

Yet, business as usual is deeply unjust and shuts the poor – which disproportionately includes Māori and Pasifika – out of the housing market, with damning social and poverty implications. Without change, unaffordable housing will be a fundamental hindrance to New Zealand creating a high-employment and dynamic economy in the future. Instead it will deepen inequality by leaving many people unable to access high-productivity jobs. This will mean more people competing for low-wage, low-skill jobs in lower-productivity places, driving down pay there further.

The way out of our crisis probably includes some support to first-time buyers, coupled with a gradual stabilisation and decline of housing prices thanks to much greater supply. But to even start on that challenging process we need to dispense with the fantasy that we can rapidly improve housing affordability without either major increases in income or a decline in house prices.

Densification of housing is also controversial. Yet to help people access opportunities in higher-productivity areas, greater densification of housing also makes sense. Affordable houses that are far from productive areas of New Zealand's major cities are only of limited benefit to helping people into good jobs.

Alongside this, much better and affordable transport links are necessary. Driverless cars may help, but could even make the problem worse.[38] Tackling our transport troubles, in a world in which we take climate change seriously, means significant investments in low-emissions public transport.

Does anyone want to buy this stuff anyway?

Without people buying what we are selling, we will quickly be out of work. The fifth big policy area is therefore to maintain high aggregate demand, enough people wanting to buy or invest, in the economy. This is especially important because we are going to produce a lot more in a world where there is still high employment but much higher productivity. There are two possible markets: New Zealand and the rest of the world. We need to target both.

Emerging economies are where consumption will grow most in the future. China and South-East Asia are still growing strongly, while Latin America is largely an untapped market for New Zealand.[39] In the longer run New Zealand's exporters looking to find new markets will need to get used to putting on their eye masks and earplugs and heading further afield: the growth markets will likely shift to places like South Asia and East Africa.[40] New Zealand was the first developed country in the world to sign a free trade deal with China. That has served us well. Yet to get to Paul and Natia's world we need to establish strong trade agreements and business connections with countries like India, Brazil and even Ethiopia. New Zealand started trying to negotiate a bilateral trade deal with India in 2010 but progress has stalled.[41] New agreements alone will not be enough. We also need to improve upon existing agreements, as we are trying to with China.

Some businesses are already pointing the way to this broadening of export markets. Miraka, a Māori-owned dairy-processing company in Mōkai, near Taupō, produces a range of milk powders, UHT milk and real-food smoothie blends. They export these value-added products to twenty-six countries. Grounded in their values of kaitiakitanga, they produce using geothermal energy. In the future, trading to diverse markets as Miraka does will be critical to keeping people employed in New Zealand.

Yet trade has become more controversial in New Zealand. This is in part because agreements like the Trans-Pacific Partnership had more to do with regulatory harmonisation – standardising the business environment across countries – than reducing tariff barriers. It is also because to get better trade access elsewhere, New Zealand has to offer something that other countries want. Trade expands the overall economic pie, but it also generates winners and losers. That is rarely sufficiently acknowledged in the rush for freer trade.

Failure to acknowledge trade agreements' costs can also mean a failure to address them. The biggest problem with trade policy globally has not been what is in the trade agreements per se (although there are many valid criticisms of this too). Instead it has been the failure to sufficiently support those who have lost out as a result of trade deals, such as businesses, and their workers, who find themselves facing stiff foreign competition in New Zealand. They need specific support – income, training and even relocation assistance – to find new opportunities for work and fulfilment.[42]

Building strong trade relationships and establishing and deepening trade agreements takes time. So does establishing programmes that can support communities that are disadvantaged by deeper trade integration. If Paul and Natia's world is what we want to aim for, we need to make both of these a priority now.

International trade, though, is not our only hope. New Zealanders living in New Zealand remain by far the biggest consumers of what we produce. If we want to live in a world of higher productivity, high employment and decent wages, we need to ensure Kiwis in New Zealand can afford to buy the goods and services we produce. To do that they need decent jobs that pay well. The Reserve Bank has one big tool to support that: monetary policy. This effectively just means setting the rate at which banks like Westpac can borrow money from them. Making it cheaper to borrow encourages greater investment and consumption in the whole economy (higher aggregate demand) and therefore tight labour markets in which firms have to try harder to attract workers. As Summers told me, 'tight labour markets are the best social programme'.[43] Running the economy like this, where businesses have to go looking for workers rather than workers looking for employers, is hugely beneficial to workers.[44] Businesses suddenly need to pay attention to retaining workers by making work better-paid and more satisfying. They are more willing to upskill existing workers and even to train unskilled workers in order to hire them. New Zealand's new Policy Targets Agreement requires the Reserve Bank to consider employment outcomes – a welcome move.

Government spending also matters to whether we can buy everything we would produce at full employment. When the economy slows, government spending on benefits and investment in things like high-speed broadband infrastructure, clean green public transport and public housing can support demand in the economy and can help keep people in work. This keeps labour markets relatively tight and firms reaching for workers, rather than the other way around. This matters because, as Summers put it to me, 'work and non-work are both addictive'.[45] If people slip out of work, it's much harder to get them back in. Better to use monetary policy and government spending to help support demand and keep employment high.

Clever macroeconomic management can make a big difference, but big structural problems like inequality can undermine demand and leave even the most brilliant boffins at the Reserve Bank and Treasury in an uphill fight. Inequality is a real threat to attaining Paul and Natia's world. The rich consume less of their income than the poor. As a result, rising inequality risks weaker overall demand in the economy. That makes it hard to keep everyone in a job, as someone needs to buy whatever we're producing. This is the spectre of secular stagnation, mentioned in Chapter Five. It can be a vicious circle. Weak demand weakens wages; this increases inequality, which in turn weakens demand further.[46] This would be bad news for workers and businesses alike.

Technological change could, if managed poorly, increase inequality and worsen the problem. An often-told tale neatly captures how: Henry Ford II is showing Walter Reuther, the leader of the United Automobile Workers, a recently automated car factory in the USA. 'Walter, how are you going to get these robots to pay your union dues?' taunted Ford. Quick as a flash, Reuther replies, 'Henry, how are you going to get them to buy your cars?' Just like in the USA, enough New Zealanders have to be employed and earning a decent income to buy the cars (and whatever else we are producing).

Tackling inequality is not easy, but there is much that can be done.[47] Broadly, measures to deal with inequality fall into two unimaginatively named camps: pre-distribution and redistribution. Pre-distribution just means trying to improve what lower-income earners actually get paid.

This includes minimum-wage legislation, stronger unions and efforts to ensure real wage competition – important given New Zealand suffers from weak competition.[48] Nobel Prize-winning economist Joseph Stiglitz has explicitly called for stronger unions to help ensure the coming technological change doesn't result in the vicious cycle described above.[49] Pre-distribution all happens *before* the taxman gets involved.

Pre-distribution tries to reduce the spread of incomes in a given business or industry so that there is not such a yawning chasm between the top earners and the bottom.[50] No easy feat, as most union leaders can tell you. In northern Europe they have a track record of doing just this. They strike compacts between unions, employers and government where, for example, workers agree to certain wage increases in exchange for greater voice in decision-making or greater R&D. These agreements mean that the difference between wages in the corner office and the shop floor isn't quite so shocking, and help reduce the likelihood of strikes.

This is not just a fanciful European idea. New Zealand has already done something very similar. The Care and Support Workers Pay Equity Settlement in 2017 was a sector-wide compact. A slight variation on the theme was the KiwiRail and Rail and Maritime Transport Union agreement, which ensured the lowest-paid workers got the largest percentage increases, thereby reducing inequality in the workforce.[51]

Redistribution, on the other hand, is basically the tax and welfare system. Consequently, it all happens *after* the taxman gets involved. New Zealand, at the time of writing, has no capital gains tax, and low top income-tax rates by international standards – a lot lower than our friends the United Kingdom, USA and Australia.[52] Taken together, some have suggested that if capital gains are accounted for, high-earning New Zealanders may pay a smaller share of their income in tax than poorer New Zealanders.[53] Put more bluntly, New Zealand is a good place to be wealthy. There may be better ways to tax – with a greater focus on problems like carbon emissions – but even on the basics our current system isn't exactly Robin Hood.

Proposals for more progressive taxation are often opposed on the grounds that taxing business or the wealthy to redistribute will hinder growth. Yet researchers at both the OECD and the IMF conclude clearly that redistribution does not necessarily lower economic growth.[54] If done well, it can even improve growth.[55]

The big issue with redistribution, as we have seen, is that it is getting less popular in New Zealand. Still, doing it at a small scale in a future in which most people receive the majority of their income from work will certainly be easier than a radical shift to the major redistribution needed for Rachel and Wei's world. Still, pre-distribution, based on a quintessentially Kiwi belief in fair remuneration for work, may be more politically popular.

The big green elephant in the room

This brings us to perhaps the most urgent policy area: reducing the environmental damage from economic growth. Growing economic output, which we need to realise is inevitable in a world of high employment and growing productivity, has historically also meant growing environmental damage. Yet we can no longer continue to disregard the state of our environment; we are up against real planetary boundaries.

A first and critical step would be to put a price on carbon, a way to require businesses to pay for emitting carbon so they emit less, that actually reflects the need to keep the world under 1.5°C of warming. The current gap between the real costs of carbon emissions and carbon prices around the world is simply massive. It is also barely closing – at current trends globally we would only be charging the real price of carbon by 2095.[56] That's seventy-five years too late. New Zealand has a particularly yawning gap. Without even including the methane that has been emitted at no cost in agriculture in New Zealand, the gap is substantially worse than almost all of Europe. The Productivity Commission has argued that for New Zealand to move to net zero emissions by 2050 the price for a tonne of carbon may need to go as high as $250.[57] In 2018 it was between $20 and $25. That is a tenfold rise in thirty-two years. For the price to be that high, our policies would have to be much tougher on emissions.

We don't yet have all the technology we need to reach net zero emissions, particularly in the agricultural sector. A higher carbon price would help spur innovators to develop these technologies and prompt businesses to adopt the clean technologies that we do have.

Other approaches, like a carbon tax, could have a similar effect. Both a carbon tax and carbon pricing in a cap-and-trade system (where businesses buy from the government the right to emit carbon) would generate money for the government. What to do with that money? One potentially game-changing option is to give it back equally to every New Zealander as a dividend.[58] That would be progressive – those who use less than average amounts of carbon (generally poorer households) would receive substantially more as a dividend than they would pay in carbon taxes. Crucially, the simplicity and transparency of it would also mean many people might like the idea of raising the price of carbon further; after all, the higher the price, the bigger their dividend.

Substantially more government R&D investment in clean technology is also a must to accelerate the transition and to address the emissions from our agriculture sector.[59] New Zealand also needs to look at incentives to shift to electric vehicles, used to significant effect in Norway, and major reforestation programmes. The longer we wait to get started the more costly the required carbon transition will be.[60]

If only climate change were our only environmental problem. In other areas like water and air quality, where higher production risks greater pollution, we need to put in place regulations that actually confront the magnitude of the problem. And we may need to compensate the losers from tighter environmental restrictions, to smooth the way and avoid further delays.

In every other major policy area we have managed something similar to what will be required in the transition to Paul and Natia's world before. But New Zealand has never before managed to grow the economy while shrinking our environmental damage. Germany and Denmark, with their strong policy commitments to renewable energy and electric vehicles, have had the most success in moving in this direction. New Zealand urgently needs similarly strong, clear policy direction if it is to decouple economic growth from environmental damage.

What happens if we don't reach Paul and Natia's world?

It will be challenging to create Paul and Natia's New Zealand, in which, despite major technological change, we maintain high employment with decent jobs and avoid runaway climate change. What could go wrong? And what would New Zealand look like if we try to achieve this, but fail?

There are, unfortunately, many ways to fail. And without proactive choices and policies like those described above we are heading for failure. First, we might fail by not focusing enough on market-creating innovation, not sufficiently combatting inequality, and not building access to growing markets. A second possibility is that we fail to satisfactorily improve our welfare and education system. A third failure might be in urban policy and transport. Those in lower-income neighbourhoods would simply not able to afford to move – or even easily commute – into higher-productivity parts of cities to access better opportunities.

If we fail in these areas, New Zealand is likely to see high and rising inequality. Many people will not be able to access high-income jobs in higher-productivity sectors as a result of their educational background, the limited number of well-paid jobs due to lack of demand at home and abroad, or the unaffordability of housing near productive areas. This heavily divided society would leave all those not able to access high-paying jobs competing for the same set of low-pay jobs in the less productive sectors of the economy. The sheer number of people competing for these jobs is likely to mean wages stay low for them. The ranks of the working poor would grow.

Paul and Natia's world is one in which people can make decent money providing personalised services to each other. High inequality, however, would mean most people will not be able to pay for personalised services. We will instead be in a world in which a few people participate in high-skill, high-pay sectors like artificial-intelligence design or business management, while most others scrabble, at low pay rates, to provide the wealthy few with personalised services.

For the reasons outlined in Chapter Four, even in this scenario, long-term unemployment may not be stratospherically high, but it could easily hit levels we have only seen during recessions. People will be so desperate for work they will find some way to augment their income. This work on the margin, however, is likely to be unfulfilling and lowly paid – even if it involves engagement with technology. As Nicholas Carr argues in his book *The Glass Cage*, work that relies heavily on automated machines can dull our brains, chip away at our expertise, and ultimately de-skill us and make it harder to learn.[61] That can further trap people in low-skill, low-fulfilment work. Underemployment is likely to rise, with people picking up what work they can through gig-economy platforms like Fiverr. Some people will also give up looking for work altogether.

To be very clear: if we fail to take proactive choices to shape our future, it is this kind of future we could be drifting towards.

Even if technological change is at the slower and lower end of the spectrum, without addressing these big challenges the poor are still most likely to be hard hit. They may be more susceptible to the automation that does occur, given many work in lower-skilled occupations; they have less savings to draw on in the event they suffer a period without employment or need to retrain; and they are less likely to live in places where higher-productivity jobs might be growing.

This already bleak vision of New Zealand does not yet include environmental damage. But we may also fail to decouple what growth we do achieve from environmental damage. If the failure to decouple greenhouse gas emissions was global, the consequences would be severe. The Royal Society of New Zealand provides a detailed description of a high-carbon world for Aotearoa.[62] It would mean temperature increases of another 2.5°C to 5°C this century, and sea level rises of up to 1.1 metres, or even more. Even with just a 0.8-metre sea-level rise, what are currently 1-in-100-year high tides would become an almost daily occurrence. This would, for example, swamp much of Mission Bay and Kohimarama in Auckland every day. Large areas of coastal New Zealand would be eroded and buildings, developments and lives lost. Whole suburbs of major New Zealand cities would become uninhabitable.

Heavy rainfall and river floods would also be much more common – and expensive. Flood damage in New Zealand in 2004 alone cost $380 million.[63] The flood that ripped through Edgecumbe in the Bay of Plenty in 2017 was so severe that a year later seventy-eight homes were still unoccupied.[64] Yet two-thirds of New Zealand's population live in flood-prone areas. Something like Edgecumbe's harrowing experience of displaced families and uninhabitable houses would become a real risk for most New Zealanders.

In drier parts of the country, on the other hand, severe drought would threaten the agricultural sector. Even assuming only moderate warming, droughts are predicted to become twice as common in Otago and Canterbury by as early as 2040. In a high-carbon world, fire risk would also sharply increase. Māori would be among the most vulnerable to these changes: declines in freshwater availability for many rural Māori settlements could also disrupt lives and investments and force communities to move.[65]

Trying to reach Paul and Natia's world but failing on the environmental dimension would be disastrous.[66]

Creating Rachel and Wei's world

Pursuing a future in which technological progress is combined with high employment is not the only possibility. Some would prefer a world like Rachel and Wei's, in which many people do not work but instead receive some form of universal income. Freed from the drudgery of work, they argue, we could pursue our true passions.[67] A future in which most New Zealanders do not work is not, in my view, technologically inevitable any time soon, as the previous chapter showed. But that does not mean it is impossible. Over the long run, policy can dramatically shape society and the economy.

Let's therefore shift our focus to look at what it would take to create a labour-light future akin to Rachel and Wei's world. Instead of trying to continue keeping people in work, what would it take to free many people from work and provide an income some other way?

This would be radical and difficult, but there are policies and choices that could move New Zealand's future in this direction. Three areas stand out: hastening the invention and adoption of automation technologies, transforming the welfare system, and supporting people to find meaning outside of work. The need to address climate change would also remain, just as in Paul and Natia's world; this is picked up again in Chapter Eight.

Hastening automation

Hastening the invention of automation technologies would need to be the first big policy priority to move towards this radical future. This would mean focusing heavily on generating efficiency innovations – which are most likely to reduce the number of people needed to produce a product or service – and less on market-creating innovations. State-funded R&D efforts and incentives for private innovation would need to expand and focus on automation and artificial intelligence. New Zealand has its own automation challenges. For example, for a wealthy country we work a lot with tourists and on farms; 7 per cent of total employment is in accommodation and restaurants, while another 5 per cent is in agriculture.[68] New Zealand R&D could focus on things like automated cleaning services, artificial-intelligence-powered tourist information and digital concierges. Similarly, we would need to invest in R&D for automation on the farm: automated planting, harvesting or shearing. The state has already made a start; New Zealand's state-owned Endeavour Fund recently granted $16.8 million to Auckland University to develop robots with arms and hands to work in vineyards.[69]

Automating other parts of the economy would likely require innovations from outside New Zealand. Retail is our second-biggest sector by numbers employed. Advances in e-commerce by Amazon could some day offer serious automation of retail – especially with widespread drone delivery. Yet even in the USA, online sales are still only about 10 per cent of total retail sales.[70] This may grow, but perhaps we also like getting out of the house and picking the best apples from the shelf ourselves. Still, cashier-less physical stores like Amazon Go, in which sensors simply track what you put in your basket and charge you digitally, could also reduce the need for workers in bricks-and-mortar stores.

Automation technologies from overseas need not stop at supermarkets. Administration work in New Zealand accounts for about 9 per cent of the workforce. Developments in artificial intelligence may be able to automate many back-office services such as accounting, business processing and even call centres. Basic back-office bots – powered by artificial intelligence – have already been developed in the USA.[71] Transport and logistics represents 4 per cent of total employment in New Zealand. Driverless vehicles, over time, may be able to take over a significant share of the tasks in this sector from humans.

Yet most of these technologies are still in the early stages of development and commercialisation. There is considerable uncertainty about how well, and how quickly, they will develop even if New Zealand's government and businesses were offering big rewards and support for their development. Above all, they are still expensive. Plus, New Zealand's track record at bringing cutting-edge technology into widespread use is weak. Despite this, there are a number of things we could do to support commercialisation and adoption. The first would be to foster much closer links with innovation hubs in places like Silicon Valley, Shenzhen, Tel Aviv, Cambridge, Berlin and Dublin.[72] To adopt these technologies in New Zealand, Kiwi businesses and consumers need to understand them and see their value up close. Concerted and regular engagement from New Zealand business, universities and government through visits, collaborations and migration schemes could form part of an overall automation strategy for the country.

This could complement a deliberate effort to position New Zealand as an early adopter for a range of automation technologies through our regulatory environment. As we saw, New Zealand was chosen as the first country for drone-delivered pizza thanks in part to our permissive regulations.[73] A welcoming regulatory environment for both the technologies and major companies like Uber and Amazon could help technologies like drones, driverless vehicles and back-office bots arrive in New Zealand as a first market.[74]

This still leaves the issue of cost. Yet if a highly automated economy is the long-term goal, choices on taxes, subsidies and government procurement could also make a big difference. For example, public-transport operators in New Zealand could be required to purchase driverless vehicles in the same way cities like Los Angeles are replacing almost all their diesel buses with electric buses. Indeed, in Stockholm autonomous shuttles are already being used as part of the city's public-transport infrastructure.[75] Government back-office administrative support could also be pushed to adopt artificial-intelligence solutions to automate tasks.

In some cases this may be more costly than hiring workers but, within reason, this cost trade-off could be acceptable if giving people the opportunity to not work was considered sufficiently valuable to society. Even today, cost is not the only consideration in many public-sector procurement decisions. Other concerns like the environment, building business capability and national trade interests also matter. Enabling wider automation and allowing more New Zealanders to not work could be another value we choose to prioritise.

To encourage adoption and innovation beyond the public sector, the government could change the tax system to incentivise automation. A payroll tax, a tax on employers for the workers they hire, would be one option. At the same time tax deductions and subsidies for capital expenditure on equipment like robots or artificial-intelligence systems could be substantially increased. Remember that automated milking sheds have existed since the 1990s, but very few New Zealand farms have invested in them – primarily because they are costly. In New Zealand these incentives would also help direct the local innovation system towards delivering more widespread automation technologies.

By positioning itself globally as a supportive and commercially attractive first market for automation technologies, New Zealand could even help incentivise international researchers and innovators to focus further on automation. A market with just enough scale and government support could be an attractive proving ground for global innovators – and help nudge them toward challenges that are particularly relevant in New Zealand. To drive innovation in areas receiving inadequate research attention, the New Zealand government could offer prizes to organisations from anywhere in the world that achieve, for example, a high-quality automated sheep-shearing robot. All these efforts to influence the direction of technology would of course be more effective still if other countries took a similar approach.

To be clear, these incentives could shift business decisions away from those that would maximise economic growth. They could lead to what Daron Acemoglu terms excessive automation, whereby *less* automation would actually allow higher output.[76] Yet there are already substantial distortions under our existing taxation systems so this would not be completely new. More fundamentally, this might be a trade-off worth making if we collectively believed that a society with widespread automation and people free to pursue other activities was a better version of society. We might, in short, accept being less wealthy if it meant we could have a decent income without most people working.

There is also a risk that pushing the innovation system to focus heavily on reducing the need for people may distract from innovating to address climate change. While the changes to our economic system proposed here could go some way towards addressing climate change, depending on how much we chase economic growth, this is a real challenge, and is discussed in the concluding chapter.

Transforming the welfare system

Over time, then, there is reason to believe we could hasten automation considerably above what would happen otherwise. There remains, of course, uncertainty about just how far and how fast this would get us toward widespread automation. And even with automation accelerated, as Chapter Four described, people may stay in work by finding or inventing new tasks that have not been automated. Without income from elsewhere, we are unlikely to give up on work. This has, after all, been the tale of modern economic history. If the intention is truly to allow people not to work, we must develop a redistribution system that can provide people with a reasonable income without working. Only with this in place is there the possibility of many people choosing not to work.

Creating such a distribution system would be a massive undertaking. Nevertheless, the broad contours of the policy prescription here are well known. First, it would require the introduction of some form of UBI. Pilots are under way from Oakland, California, to Kenya. This could be supplemented by the provision of what I have called 'gift stocks', whereby each citizen is given a diversified portfolio of shares on their eighteenth birthday. This would provide dividend income as well as an asset that could be sold or traded.

How could we afford all this? Gareth Morgan and Susan Guthrie's proposal for New Zealand suggests a UBI of $11,000 per year could be financed through a flat tax of 30 per cent on all income. $11,000 is a relatively low UBI, and for this future to be attractive – and for people to elect not to work – a significantly higher UBI may be needed.[77] A wealthier New Zealand, which is probable with long-run technological progress, would be able to afford a higher UBI at the same tax rate. Yet if fewer people are working and paying income tax, tax rates may need to be higher to support a UBI. These tax revenues would also need to fund the government's purchase of gift stocks.

There is a range of possibilities for additional revenue, including higher income taxes, the introduction of payroll taxes (consistent with incentivising greater automation), higher corporate taxes, capital gains taxes, higher taxes on 'bads' like carbon emissions, or possibly wealth taxes. All this talk of higher taxes may sound alarming. These are only possibilities, each with their own very real trade-offs; which, when and how much of these might be needed would depend on the level of basic income and state of the economy.

Additionally, if government was investing significantly in creating automation technologies it would also have a legitimate claim on the returns from those investments, as Mariana Mazzucato has argued. This would reduce the tax revenue required. The pain of higher taxes would also be partly offset by the universality of the basic income. Still, a higher tax burden overall for the wealthy and for the owners of robots would almost certainly be necessary to create a world in which many people do not have to work.

Could this tax burden kill the goose that lays the golden eggs? It could. Maintaining higher tax rates without damaging business investment or innovation to the point that the tax burden undermines the productive sectors of the economy would be a tricky balance – both economically and politically. Failure to strike that balance would mean that efforts to hasten automation, the other side of the coin, could be undermined. Getting the sequencing right – not ramping up taxes too far until automation has progressed further – might help but would require delicate management. Even with this, there is a risk that higher tax rates would scare away both the investment needed to hasten automation and the talent needed in New Zealand to help realise it.

Yet it would also be wrong to dismiss it as impossible. Sweden, Denmark and Finland all built world-leading economies – including successful innovative companies from AstraZeneca to Maersk and Spotify to Ericsson – with top income-tax rates of around 60 per cent and widespread freedom of movement for those who wanted to move elsewhere in Europe. Similarly, in the era of rapid innovation between the 1920s and 1970s (discussed in Chapter Three), the USA had top marginal income-tax rates of as much as 90 per cent.[78] Moreover, at the time the USA generated thousands of technology companies including world leaders like Google and Amazon, it had the highest corporate tax rates in the OECD at 39 per cent.[79]

Finding meaning and happiness in a world without work

How do you support people to find meaning in life without work? Today people on welfare face an unsympathetic and at times downright hostile environment. Meanwhile, working hard – typically narrowly defined to exclude tasks such as housework – is lauded as a sign of excellent character. These ideas are heavily ingrained in our culture.

Rachel and Wei's world would mean a radical change in attitudes to work and welfare. There is, however, nothing in our genetic make-up that means we must think as we currently do. Societies have been organised in other ways over the centuries, and some still are. The hunter–gatherer society of Bushmen in Southern Africa did not have the modern concept of work. Similarly, the Pirahã of the Amazon do not divide life into work and leisure. In both societies hunting and gathering is seen as something fun, more akin to play. In the Bushmen's society, hunting is only done when it is necessary. Even if we define activities as work through a western lens, in Pirahã society most adults work only twenty hours a week.[80] In line with this thinking, back in 1968 the anthropologist Marshall Sahlins branded hunter–gatherers as living in the original affluent society.

Life can be very tough of course in these societies, but the real lesson is that we are not somehow genetically predetermined to work as we do or to suffer mental-health problems when we do not work. Instead our work habits and attitudes are a function of our culture and choices. For New Zealand to move through a democratic process towards a future of far less work, there would be a large role for civil society, media, business and government to help build wider societal acceptance of not working, and living on income from the state.

Stigma of not working aside, people still need things to do that provide meaning and flow. One possibility would be for both parents to actively bring up their young children – rather than sending them to childcare for large portions of their waking hours or leaving it primarily up to mothers. Indeed, the freedom not to work could radically reshape typical gender roles, with men able to be far more involved in their children's lives right from birth. Both parents could also be much more involved with their children's education even into their teenage years. Others have emphasised volunteer work, creative pursuits and far greater time for social connection.[81] Another possibility is that without the necessity to work, people would be free for greater political and civil engagement, such as political campaigning or community service.

Some people may easily find such meaningful activities. Others will need support. Today's retirees in fact already provide some pointers to finding meaning without work. One example is the Men's Shed movement, which provides workshops in local communities for primarily older men to go and work with their hands, doing woodwork, metalwork and repairs, among other things.[82] These workshops provide an opportunity to socialise and learn from other men; participants claim that they have changed or even saved their lives.[83] Men's Sheds are now the subject of academic studies looking at their benefits on everything from mental health and dementia to diabetes.

There are other similar initiatives aimed at providing meaning to people. The Columbus Idea Foundry is the USA's biggest 'makerspace', where hundreds of members of all ages pay a small fee to use its machines to make crafts, paint, weld and laser cut. It is above all a social place and even has its own communal pizza oven. Members certainly find meaning there. One member, Terry Griner, told *The Atlantic* that 'if we had a society that said "we'll cover your essentials", you can work in the shop. I think that would be utopia.'[84] Creative social places like this could be an example of how to provide meaning in a low-employment future in New Zealand.

As Chapter Six already discussed, education would also need to be radically different and focus far more on finding passions rather than finding employment. But the dramatic shifts in policy would go well beyond education. Agencies like WINZ, whose primary objective is currently to get people back into work, would also need to be reimagined. Such an approach would also have implications for other areas. Urban policy would need less emphasis on getting people access to the CBD and greater emphasis on fostering community connections. Even less obviously connected government agencies like conservation might need to broaden their approach to focus, for example, more on community participation and engagement.

These changes would be truly radical, but are not impossible in the long run. The question is whether they are desirable.

What happens if we fall short of Rachel and Wei's world?

Aiming to create Rachel and Wei's world would be, on most counts, more radical and risky than aiming to create Paul and Natia's world. There is far greater uncertainty around how policies to shape this future would play out, because we have never done anything quite like it before. How could this go wrong? And what might the future look like if efforts to reach Rachel and Wei's world failed?

The first big thing that could fail is redistribution and taxation. The world has no major historical track record of succeeding at redistribution at this scale. In some places large, almost universal financial support from the government is commonplace. In the United Arab Emirates, among many other payments, the state gives Emirati men about US$19,000 when they marry an Emirati woman.[85] In Alaska, citizens receive a modest monthly dividend from the state's oil revenues. Yet these transfers are financed through mineral or petroleum wealth. Financing a UBI through taxation on the regular economy is much more difficult.

First, it is politically more difficult to build support for such taxation, although the universality of a UBI may be more popular than high taxation for spending that solely goes to the poor. Second, high levels of taxation risk dampening economic output and innovation by diminishing the returns to investment. That could reduce the prospects of sufficient automation and shrink the overall tax take by decreasing the size of the economy. Another big failure might be that people struggle in finding meaning and purpose outside of work more than we might expect – and get less support than they need.

A world in which we did hasten automation but failed to redistribute sufficiently or support people to find meaning elsewhere would be ugly. Many people would be out of work and, lacking sufficient alternative income, would face severe financial hardship – difficulty making rent, buying groceries or paying for medical care. It would be the worst of both worlds: no job, no income support and no purpose. Homelessness and poverty would surely rise. Moreover, even the more financially secure may suffer from a loss of meaning and purpose in life. Rather than finding new social groups and meaning in volunteer activities, they may instead suffer from social isolation.

Could the real constraint to a world of much less work simply be our choices?

There's another big risk to aiming for a labour-light world: technological progress may be slower and less comprehensive than hoped, despite the best efforts of the government to incentivise further automation and rapid productivity growth. This could mean that the national economic pie would be smaller. What could be redistributed through a UBI would therefore be smaller too. What would this mean for a labour-light world?

We tend to think about a future without work by asking when technology will be advanced enough for us to stop working. The implicit assumption is often that once technology does nearly everything we do today, we will stop working. However, technology has replaced nearly everything we were doing on a given 'today' multiple times in the past – yet we are still working.

Agriculture is the sector in which this is most obvious. Agricultural employment has declined dramatically in all rich countries. In some the automation of agriculture happened fast. In the 1940s about 50 per cent of working Finns were in agriculture. Just forty years later less than 10 per cent were.[86] In South Korea agriculture declined from 50 per cent of employment in 1970 to just 11 per cent by 1997.[87] That's ten million farm workers leaving their jobs in just twenty-seven years.

In all these cases the vast majority of people continued to have enough to eat; in fact, most had more to eat. Agriculture productivity growth was so remarkable that a plethora of tasks previously done by humans were now completed by machines. Agriculture was in effect almost entirely automated. In South Korea, 40 per cent of total employment was eliminated in just twenty-seven years.

Yet those workers in South Korea found new things to do; new tasks were invented, new ways to make more money were created – many of which the average Korean farmer in 1970 would have struggled to imagine. In fact labour-force participation in South Korea actually rose. Automation, for many of the reasons discussed in this book, did not lead to unemployment.

In theory, at least, South Korea could have chosen differently. They could have collectively decided that the ten million farm workers who'd had their jobs automated between 1970 and 1997 could stop working. Instead as each of those ten million workers left agriculture, the South Korean government could have provided them with a basic income from taxation on the economy as a whole.

This would have meant much less economic growth, as those ten million workers would not have produced anything. It also would have meant South Korea could not have had as much of many of the things they enjoy today that require human labour, such as smartphones, music videos or dining experiences. Nonetheless, had they been broadly happy with living standards at late-1970s levels and found a way to redistribute income, those ten million Koreans could have stopped working. Yet they did not collectively choose to. Moreover, they did not put in place a redistribution system that would have allowed individuals to easily make the choice to work less for themselves.

Instead, like everyone else, they wanted to be richer and to enjoy new experiences such as regularly eating out or going to the movies.

These experiences still required human labour and so South Koreans kept working. There are good reasons for these choices, of course. Koreans did not just want to be wealthier to enjoy more dinner and movie dates. Being wealthier meant they could do countless things such as buying healthier and tastier food, accessing better health care to combat everything from cancer to toothache, travelling to experience other cultures and countries, getting a better education (perhaps even abroad), using technologies like mobile phones to stay in touch with loved ones, and helping out struggling members of the wider family.

All this suggests a rarely acknowledged lesson about the possibility in New Zealand of a future with far less work: it may be our choices and policies, not technological limits, that fundamentally stop us. We continue to make choices, both individually and as a society, that prioritise continued growth and income over working less.

Automation of many of our day-to-day tasks has happened before. Yet we have not stopped working, bar some small reductions in average hours per year. Instead individual New Zealanders have found new things that are not automated to work on and earn income from, in part because we want to consume more and experience new things, and in part, some might argue, because without a collective choice to provide some other form of income through redistribution, it is difficult for individuals to choose to work less. We have not collectively tried to establish any alternative source of income like a UBI, quite possibly because people stopping work would make us poorer as a nation.

With technological change highly unlikely to be so sweeping or rapid that no one can come up with new tasks for humans to do, the choice is likely to remain ours for a lot longer yet.

Seen from a different angle, this implies that even if efforts to hasten automation in pursuit of Rachel and Wei's world were less successful than hoped, we could, at least in theory, still move toward a world of far less work over time. This might take longer, and the trade-off, of course, would be that we would do so with less overall income to share around. We would be less well-off than if we attempt to introduce a UBI and reduce the need to work from a position of higher levels of productivity and income. That would likely make it both far less attractive to many and far harder to implement in practice.

It would certainly be a radical trade-off, but it is worth being clear about the prospects for rising incomes in such a society. Suppose that technological progress was nearly as dramatic as some claim, and in New Zealand 40 per cent of the work force, about 1.1 million people, was displaced by machines between 2030 and 2050. Imagine that as each of those 1.1 million workers was displaced they did not find new work and instead exited the labour force and received some form of basic state income. By close to 2050 the level of output would be lower than if those people had tried to find new work. After all, almost 1.1 million would no longer be contributing to economic output.

It is unlikely, however, that New Zealand would never grow again. We would be substantially poorer than we would have been, because fewer people would be contributing to the economy. We would have dramatically shrunk the labour force, while leaving the now smaller labour force to generate income for everyone. This is the same dynamic that happens as the baby boomers generation retires and the smaller working-age population has to support them. But as long as the economy continued to experience productivity growth, economic growth would eventually also continue, just from a lower level.

The real question for those who argue in favour of a future with far less work is not when will the technology be good enough to automate everything we do today? Instead it is the following: to what degree are we willing to trade off being poorer, giving up some services that machines cannot yet provide, and growing more slowly in order to build a society in which people do not work? Sitting behind this is an even simpler question: how much is enough?

The idea that we might simply collectively choose to help a large chunk of the population to not work and instead live primarily off an income from the state – even if that made the country collectively poorer – might sound a bit crazy. Except for the fact we have already done exactly that in New Zealand. In 1938 the First Labour Government introduced a universal income for everyone over the age of sixty-five. These days it is called superannuation. That still exists today, is universal above the age of sixty-five, and for a couple is set to 66 per cent of the average weekly take-home wage in New Zealand. In 2018 that meant a retired couple received a basic income of just over $32,000 a year.

To be clear, the choice to introduce a universal income above sixty-five *did and does* make us economically poorer. Many New Zealanders do retire at age sixty-five, largely because of superannuation income, but also because it is an accepted and supported social norm. Yet we would be richer as a country – we would produce more – if all those workers kept going full time until they were seventy-five or older. The fact most people stop work and receive a state income at sixty-five means we have a lower GDP and lower tax revenues. But we have simply chosen as a country that the trade-off of being poorer as a nation is worth it to allow people to stop working at sixty-five. That is fundamentally a choice.

Extending a universal income to all adults would of course be a dramatically larger trade-off and, as has been discussed, would likely require higher taxes. Yet we are also already much richer than we were in 1938 – and by the time we might move toward any scheme like this will likely be richer still. Back in 1938, the economy was a lot less automated. We were a lot less productive. And we were a lot poorer. Yet we still made the choice to provide a universal income to everyone above sixty-five because we thought that in a good society once you hit sixty-five you should not have to work – even though it made us poorer as a country. We collectively made another major choice in 1936 to limit work further, even if it meant we produced less as a country – we created weekends.[88]

A similar choice and trade-off could be on offer in the future. Technological progress helps improve the terms of the trade-off. The richer we are thanks to higher automation and productivity, the higher the universal income can be. Yet if we could provide a universal income to the over-sixty-fives back in 1938, it is not necessarily insufficient technological advancement that will stop us. It may instead be whether collectively we like the trade-off of less wealth for more leisure time. Once again: how much is enough?

The day-to-day difficulty of shaping the future – and how to start

The idea of choice in this chapter has been used, with an eye to the long term, to demonstrate that our future is not somehow already technologically or economically predetermined. What we choose to do as a society, as individuals, as businesses and via our government can have a profound impact on the future we create – particularly over the long run. If New Zealand is a mighty waka heading across the open ocean into the future, we five million Kiwis are not just sitting in it; we can paddle and navigate it. We may well be battered on the seas of the global economy, but if we can establish a clear sense of where we want to go – and paddle in the same direction – we have a decent chance of getting there.

Still, the currents of the global economy are strong. As discussed, there is a risk, particularly in shifting to a labour-light future in New Zealand, that too dramatic a change could scare away foreign capital, which is an important support to our economy, or push our best minds to leave New Zealand. Gradual change, adjusting and harnessing the waves of the global economy even if it means a more circuitous route, would certainly be wise.

Even setting a direction will not be easy. Taking collective decisions through our democratic political system is difficult. Political polarisation in New Zealand, while far less pronounced than in other parts of the world, can make it difficult to face long-term collective challenges such as the future of work. New Zealand's short-term political cycles further sharpen the challenge. Yet, as Jonathan Boston has outlined in his timely book *Safeguarding the Future*, there are things we can do to strengthen our capacity to tackle these challenges.[89] The introduction in 1938 of a universal pension for anyone over sixty-five suggests that it is not entirely beyond us to make big collective decisions about work, income and what kind of a society we want to live in.

Perhaps above all, to make important collective decisions about New Zealand for the long term we have to remember that our own agency as individuals and businesses and through government is greater than we often think. Cynicism or resignation about our ability to shape the future, a sense that the waka can do no more than drift with the technological and economic currents, is both mistaken and perilous. Worse, it risks becoming a self-fulfilling prophecy; if we stop paddling we really will be at the mercy of global currents. And of course, failing to choose a direction at all is a choice itself, just not a good one.

There is still, and should still be, a lot of debate on the direction we should take – both between the broad futures outlined here and on the details of whichever broad direction we favour. Even if it takes time for New Zealand to set a direction for the future of work, it would already be progress for us to remember that the future is in our hands and to begin the debate about what kind of a future we want to create. This book aims to be a small contribution to that effort.

Eight
Making choices

> The best way to predict the future is to invent it.
> – Alan Kay, EDUCOM Bulletin

The future of work in New Zealand is not yet written. Instead it is up to us as individuals, communities, iwi and businesses, and to our government, to shape it.

This book has underscored just how much policies and choices shape the future of work – something we too often forget. The good news, as the first part of the book argued, is that we do not need to panic. More of the working-age population than ever before is in work in New Zealand and there are few reliable signs that widespread technological unemployment is on the horizon. As Chapter Two showed, new technologies can take a long time to make it into productive use. And even if major new technologies (including automation technologies) do arrive more widely, there will be a number of effects (productivity, upgrading and new tasks) that will help workers and jobs. This means we have time to shape the future. That's critical, because shaping it takes time.

The bad news is that if we do not make purposeful choices to shape the future of work, our default policies, business practices and individual attitudes will shape it anyway – just not necessarily in ways we want. That could see us eventually drift towards a world with growing inequality, an excess of efficiency innovation that replaces workers, and not enough people able to afford to buy what we are producing. This would see good jobs harder to come by and increased underemployment; work for many would be uncertain and unfulfilling; and in the absence of an alternative source of income, poverty and inequality would rise. Climate change, of course, would likely make this world much worse still. The wise move is to start now in debating and shaping the kind of future we might actually want.

Making big decisions to shape the future of work in New Zealand will be tough. Getting from millions of individual opinions to a direction for the government and country is a fraught process. That is, though, the purpose of our democratic institutions and processes. And the first step is to have a debate on where we want to go. This book has sketched out two very different possible futures for New Zealand: Paul and Natia's world, where work still plays a central role in life, and Rachel and Wei's world of basic incomes, where work plays only a small part. These are broad sketches of possible futures. In the long run we can shape our future in either direction. And of course, these are representations that sit on a continuum of how much we might work. We could shape a different version of these broad futures.

I have tried to describe each future and its challenges without mentioning my own preferences. I believe above all that we need to focus clearly on the fact we have real agency to shape our future– no matter the direction we favour. Yet I do have a view; I believe that a future in which work is an important part of life for most people is by far preferable. In other words, I think we should aim for some version of Paul and Natia's world. That does not mean we have to work the exact number of hours per week that we do now (nor that we all have to live in Miramar, eat local blueberries, have two kids and install sensors in our bathrooms). Shorter work days (as in Swedish experiments), or even eventually a three-day weekend, could improve our well-being by allowing more time for family, hobbies and friends.[1] Yet I believe that a more radical shift toward a society like Rachel and Wei's, in which we work far less, would not be wise. That may strike some as lacking in imagination. By way of conclusion, let me explain my view.[2]

First, a New Zealand in which most people still work will almost certainly be a wealthier New Zealand overall than one where many people do not (or work very little). Fewer people working productively means we, as a country, will produce less and many individuals will be poorer than they could have been. That should not be underestimated; income is still an important contributor to well-being and life satisfaction. As discussed in Chapter Six, studies show life satisfaction increases with income, although not always quite as we expect. This is especially true for poor citizens, but it's also true for richer people – although it takes more money to increase their life satisfaction by the same amount.[3] Still, this should give us some pause about accepting a poorer New Zealand. Many studies on life satisfaction focus primarily on the income of individuals within a given country – the difference in life satisfaction between a poor Kiwi and a rich Kiwi. Yet our choices about the future of work will also affect how rich New Zealand as a whole is, and that too makes a big difference to the life satisfaction of New Zealanders.

Beyond the direct benefit of a higher pay cheque for citizens, being wealthier and having a higher income as a country allows communities, businesses and government to do an extraordinary array of valuable things. This is thanks to greater charitable giving, higher profits and greater government revenues. A higher-income New Zealand would allow the government to spend more on schools, hospitals, medicine, sports programmes, the arts and music to name just a few possibilities. The outcry when Pharmac refuses to pay for a specialist new cancer drug is directly tied to what we can afford as a country. In a world with little work, the limits to what we can afford will be greater than they could be in a world of work.

A higher-income New Zealand would also mean government and business alike would have more to spend on innovation to tackle tough unsolved problems: cures and care for dementia, new approaches to address youth mental health, or new technologies to reduce greenhouse gas emissions from agriculture. In a higher-income New Zealand, community groups and non-profit organisations will be more able to find the funds to help the needy or disabled, to revitalise our public spaces and to support cultural exchange and understanding between New Zealanders of different backgrounds. Whatever your programme, it has a better chance on average of finding funding and support in a higher-income New Zealand. The evidence on the importance of national income is clear: people in richer countries are generally a lot more satisfied with life than people in poorer countries. Moreover, people with better health and life expectancy, often enabled by a country's wealth, are more satisfied too.[4]

There are trade-offs, of course. In an exclusively income-focused system the drive for wealth can create the same problems we later seek to address through spending. Moreover, most people retire at age sixty-five, and, although this makes us poorer as a country, it is a widely accepted trade-off. Yet, within reason, I would rather live in a higher-income New Zealand than a lower-income one. Aiming for a world in which most people continue to work gives us by far the best chance at that. At the very least, we should not lightly choose a path in which we will be poorer as a country.

There are also other benefits to a world in which we work. One is particularly important: human meaning. In Chapter Six I discussed how devastating the mental-health impacts of unemployment can be. I have seen up close in my own life how a lack of work can leave people struggling. Research suggests that the mental-health effects of underemployment can also be severe.[5] Work is not always the answer, but it is often a hugely important source of meaning. Its absence can take a heavy toll.

Of course in a society with different attitudes to work, the downsides of not working may not be as severe. Yet working is not just about avoiding problems; it can be a source of positive and life-affirming experiences. The satisfaction of getting a task done – whether that is the last coat of paint on a new house, a well-delivered presentation to a tough client, or the delivery of a newborn baby – can be tremendous.

It is at work where we most often find ourselves in a state of flow:

that feeling of being totally immersed in an activity, energised by it and enjoying the process. Flow seems to me sufficiently important that we should care deeply about the opportunities we have to experience it. This is most likely to occur when we are doing something challenging that demands the best of our abilities. That can happen in leisure time as well as at work. Yet to reach a state of flow, we often have to be challenged and be committed to trying to achieve a goal.

Speaking from personal experience, committing to take on a challenging task in my leisure time can be difficult. Often the desire to blob in front of the TV wins out. Yet when I do engage in an active challenge that uses my skills – whether through playing football, cooking, or just finally fixing the damn skylight – I sometimes find myself fully immersed, with time flying by. The trouble is, even if taking on something challenging and skilful might provide real meaning and flow, we often find it hard to commit. This is why the work environment matters. Work helps us make those commitments because in many cases we have to do what our boss asks – or we need the money. That can be the nudge that gets us to take on a challenge and gives us a chance at finding flow.

Without work, flow is indeed harder to come by. Studies of retirees suggest that retirement tends to come with fewer flow experiences.[6] It also tends to come with a lot more television. I like watching television – sport in particular – as much as the next person. But lengthy stints in front of the TV, no matter how exciting the match, are not a recipe for flow.

Work is of course a deeply imperfect source of flow, meaning and satisfaction. Many people are, at least sometimes, frustrated or unhappy at work. I have been at times too. A glance at the sometimes haggard looks on the faces of my fellow commuters suggests I am not alone in that. Work may be even tougher for people in difficult physical jobs like farming, construction or road works. Jobs that are neither challenging nor social – of which there are too many – can also be debilitating. Then there are people in precarious jobs, facing uncertainty every day about whether they will have enough hours to pay their rent and put food on the table. Others are excluded or marginalised from work altogether. There are many problems and imperfections with the world of work.

Yet to me the best response to this is to find ways to make work more meaningful, inclusive and interesting, with greater opportunities for flow. Technology will help by reducing the need for humans to work on some menial, repetitive or dangerous tasks.[7] As Chapter Five has emphasised, there are also plenty of policy choices – notably in education – that increase the opportunities for meaningful, well-paid work. Other choices on wage setting and workers' rights can help improve some of the worst aspects of work. This remains a momentous task. Yet it also strikes me as a far better use of our energies and creativity than finding ways to make a world of very little work function.

This book has emphasised strongly the extent to which we can shape the future. That is critical. But of course that does not mean we know what the exact outcomes of our efforts to do so will be. The greater uncertainty about how a labour-light future – Rachel and Wei's world – might play out is the second big reason I favour a world with work.

There is considerable uncertainty about how fast and how far technological progress will actually go – even with the best long-term policies in place to hasten automation. In the long run technological progress is what makes us more productive and better off. As Chapter Seven highlighted, we could move toward a world with less work no matter the level of technology we have reached. However, the more technologically advanced we are, the bigger the national pie and the easier it would be to make such a shift. Yet, as Chapter Three suggested, the evidence that we are about to see unprecedented rates of technological progress and productivity growth is still weak. There is time for this to change – this book has placed any labour-light future *at least* twenty-five years into the future. Still, the current struggles with productivity growth, and the inherent uncertainty about the path of technological progress, suggest that we could attempt to move to a labour-light world only to find we end up doing so at a lower level of technology, and thus income, than we expect. That would mean a smaller pie to share through a basic income, and a future that is a lot less attractive than we had hoped.

There is a second major uncertainty at the heart of the idea of a labour-light world: whether we can successfully redistribute in the way required. Many policy choices are difficult, but redistribution is particularly hard, because those who would be taxed tend to be both influential and against redistribution. A world like Rachel and Wei's would require much more redistribution than we currently have. Some countries have managed to reverse rising inequality, but usually only a little. Indeed, history provides a cautionary tale; we have never managed such large-scale redistribution before without violence. Redistribution is still ultimately a democratic choice. My concern is simply that when push came to shove we would not choose it. It is possible that if automation swept the country, with robot owners turfing people out of work and eliminating their prospects of finding new jobs, there might suddenly be the political support and momentum to tax and redistribute at the scale required. Yet, as this book has argued, such technological change is far from inevitable. We could aim to hasten automation in this direction and use these changes to build support for a strong universal basic income. The trouble is that we might choose to hasten automation but fail to follow it up with the tougher choice of large-scale redistribution. Unfortunately, that would just lead us to rapidly deepening poverty and inequality.[8]

To be clear, there is also uncertainty about our ability to generate good jobs for people in the face of technological change. Indeed, we are not on track to create Paul and Natia's world right now either. More than anything we are drifting and failing to shape the future proactively. This risks a world of growing inequality and underemployment, where too much work lacks meaning and is poorly paid. Nevertheless, we have at least done some version of what would be required for Paul and Natia's world before. Whether in New Zealand or elsewhere, we have revamped our education system for the industrial revolution, we have built cities that have allowed millions of people to access highly productive jobs, and we have provided social safety nets that have brought more people than ever basic security in childhood and old age as well as when misfortune strikes. Prudent management of the economy by successive finance ministers and Reserve Bank governors in New Zealand has meant that on average we have had enough people wanting to buy the things we produce to keep most people employed. None of this has been easy or smooth; in fact it has often been messy, uneven and costly. It is likely to be so again. Nevertheless, while the challenges in each of these areas ahead may be greater than before, the playbook is not entirely unfamiliar.

There is, however, a notable exception to this track record. In New Zealand we have never managed to consistently grow without increasing greenhouse gas emissions. For many who advocate for a labour-light future, climate change is the one compelling reason to make such a seismic change to our society and economy. If the imperative to maintain full employment was removed, the need to keep increasing consumption would be weakened too. (Currently we need to keep consumption growing in order to keep employment high, because technological change gradually allows us to produce the same amount with fewer people.) In this light, Rachel and Wei's world may be appealing because high employment would not be a priority; it would open up the possibility of lower economic growth and less consumption, which could take pressure off the environment.

Yet there is a big problem with this as a way of saving the planet: it may not work. It would indeed take some pressure off the environment but, given how much we've emitted already, we need far more than that. Drastic action not just to lower growth but to actually contract global GDP by 10 per cent in the next twenty years would – given existing levels of energy efficiency and carbon intensity – reduce emissions by exactly 10 per cent. Reducing GDP by 10 per cent would be an economic contraction four times bigger than the 2007–8 global-financial-crisis recession. And it would reduce emissions from the current level of 32 billion tonnes of carbon dioxide per year to 29 billion tonnes per year.[9] Yet according to the Intergovernmental Panel on Climate Change, the United Nations scientific group on climate change, the world needs to reduce carbon dioxide emissions to 20 billion tonnes per year within two decades in order to limit warming to 2°C. An economic contraction four times the size of the world-order-shaking recession of the late 2000s would be nowhere near enough to deal with climate change. Bigger permanent economic contractions are extraordinarily difficult to imagine – the implications for spending on education, health and retirement alone would be shocking. All this means that we desperately need massive reductions in greenhouse gas emissions per dollar of economic output. We need to be dramatically more efficient.

So, how do we best get such massive increases in efficiency? Behaviour changes by consumers to buy green could make a big difference. But to really tackle the problem requires new green technological innovations and widespread investment in installing existing clean technologies. After all, for most consumers to choose greener options, those options have to be at least available, ideally at broadly reasonable prices.

Continuing to aim for an economy of work seems to me to offer a far better chance of both new green innovations and widespread investment in installing existing options. First, an economy operating at close to full employment – and thus higher economic output – will likely offer greater rewards for innovation in general. To get the innovation we need, it has to pay off for those investing in it. More pertinently, if we divert much of the innovation system to focus on automating work so as to hasten toward a labour-light economy, we reduce its focus on reducing greenhouse gases. In addition, we need major investment in adopting and installing existing clean technologies like green housing, renewable energy and green transportation. An economy that is operating at close to full employment – and thus probably growing – is far more likely to generate major investments in installing clean technologies than an economy that is not. Finally, making these kinds of investments is in any case likely to result in significant job creation. Of course, for a high-employment economy to generate this green renewal there must be, as Chapter Seven underlined, a serious carbon price or equivalent regulations.

Climate change is a wicked problem. Aiming for a world with little work does not look to me like the most effective way of solving it. We should not make a fetish of economic growth, and consuming less would help reduce emissions, but decoupling economic activity from environmental damage, despite slow progress so far, is the real key. There are a few green shoots of hope on decoupling. Both Germany and Denmark appear to have decoupled their emissions from growth in absolute terms, even when taking into account their imports.[10] That is a remarkable achievement. Yet to truly address climate change we need many more countries to do the same – and continue to do it year on year – despite being poorer than Germany or Denmark. That requires much stronger policy support to push decoupling further and faster, but it does not necessitate a world with very little work.

My vision of the good life, in short, sees us living in a wealthier society where most New Zealanders work. This work in turn provides them with a decent income, social connections and, at least reasonably regularly, a sense of purpose and flow. Through the innovation and investment this work supports, as well as a far more meaningful carbon price, I also envisage a New Zealand that massively reduces its emissions so as to combat climate change.

Could I be wrong about some or even all of this? Yes. Prediction is still difficult, especially about the future. Could other people look at the evidence and draw a different conclusion about what kind of future is desirable for New Zealand? Absolutely, and as this book has underscored, we could pursue a very different future if we thought it preferable.

Yet, there are some things we need to do regardless of which future we prefer. Runaway inequality, in addition to being deeply unjust, would destabilise both a future of continued work and a labour-light future by making it unlikely we could afford to buy everything the economy produces. In the first case, this would make it harder to keep people in work. In the second case it would mean lower production, making the pie to be redistributed smaller. Someone still needs to buy what is made, no matter who produces it. The bottom line is that too much inequality could beget even worse inequality and economic troubles, taking New Zealand further still from its egalitarian roots. Keeping inequality in check is a 'no-regret move' in shaping the future of work. There is a second big no-regret move: seriously tackling climate change. In the long run there may be real differences between green growth versus de-growth (shrinking the economy) as the best way to limit climate change. But that should not get in the way of the basics: getting an effective price on carbon (and equivalent emissions) that reflects the real cost of carbon to the world. We are absolutely nowhere near that right now.[11] The third no-regret move is to strengthen, not weaken, social safety nets. In a future of continued work, we are likely to see a lot more people having to change jobs and retrain; they need real support to be able to do that. And if we move toward a future of far less work, stronger safety nets would be a first step on the way toward the kind of redistribution needed for that future to succeed.

When the facts change, I change my mind. What do you do, sir?[12]

As the world changes, the evidence that has led me to conclude a world of work is preferable might change. The trade-offs involved in creating either of the broad futures outlined here could be very different. We should frequently revisit the evidence to see if the trade-offs and possibilities have altered and whether a more radical change to our society and economy might be more feasible, wiser or simply necessary. What specifically should we look for?

First, we should keep a clear-eyed perspective on the progress of technology. This is not something we've been good at in the past – we get swept away in the hype. Yet if widespread technological unemployment does start to look inevitable, we would need to grapple hard with creating a successful world without work. What should we check on to see if it were starting to look inevitable? Given our tendency to get caught up in the excitement, it is worth simply listing these:

- machines replacing humans *at large scale* in tasks and jobs where people have typically been valued primarily for the human touch
- fewer new tasks and jobs being invented for humans, with machines and computers instead quickly taking on the bulk of the work for any new idea, project or business
- significant and sustained increases in productivity growth in rich countries and, in particular, very high productivity growth in important economic sectors in New Zealand
- high and rising unemployment (and underemployment) that does not decrease significantly in better economic times despite sound economic management
- steep declines in labour-force participation.

Remember, today none of these are happening in New Zealand.

Second, we need to see whether Paul and Natia's world of continued work despite technological change can function as hoped. Again, if it did not, it could be wise to change our direction and try to shape a future in which work plays a less significant role. Three big things, if they came to pass, would indicate to me that it may be time to think again:

- a failure to make progress in strong decoupling in New Zealand and in other countries around the world despite appropriate carbon pricing or carbon taxes
- rapidly deepening income inequality due to further declines in the labour share as well as to a growing gap between salaries for those at the bottom and those at the top
- major declines in measures of satisfaction or flow at work.

In my view these should give us serious pause and prompt us to ask two things. First, are we really doing what is needed to make a world of work a success in a time of technological change? Second, if we are and we still have these problems, is it time to look at a different social model altogether, in which work may play a less central role?

I am unbelievably lucky to be a Kiwi. New Zealand is an extraordinary place: stunningly beautiful, culturally rich, creative and down-to-earth. At our best we in New Zealand have a deep commitment to fairness and caring for others. Buttressed by this, we are also increasingly doing the hard work of grappling with our complicated past. Despite these strengths, we cannot afford to sleepwalk into the future. Doing so risks eroding what we have and leaving our children and grandchildren a New Zealand that has lost its magic.

Our choices today and in the coming years matter deeply to the future of work in New Zealand. We need to take up the challenge of actively shaping the future. That will be hard work. It should also start with a debate about the kind of future we want – ideally one that is grounded in the realities of technological change and adoption, a recognition of the power of the economy to adjust to changes, and an embrace of our own influence over the future. My own view of the good life is clear.

What is yours?

Endnotes

Epigraphs (in chapter order)

May Davis, *May Davis: Her Story*, Nelson, 1990, p.21; quoted in David Rotman, 'Who will own the robots?', *Technology Review*, 16 June 2015, www.technologyreview.com/s/538401/who-will-own-the-robots (accessed 11 February 2019); Fredrik Erixon and Björn Weigel, *The Innovation Illusion: How So Little Is Created by So Many Working So Hard*, Yale University Press, New Haven, 2016, p.185; Tim Harford, 'We are still waiting for the robot revolution', *Financial Times*, 30 June 2017, www.ft.com/content/4423a404-5c0f-11e7-9bc8-8055f264aa8b (accessed 18 February 2019); quoted in Melvin A. Benarde, *Our Precious Habitat*, W.W. Norton & Company, New York, 1973, p.v; Elon Musk (@elonmusk), Twitter, 14 April 2018, https://twitter.com/elonmusk/status/984882630947753984?lang=en (accessed 11 February 2019); Erik Brynjolfsson and Andrew McAfee, 'Will Humans Go the Way of Horses? Labor in the Second Machine Age', *Foreign Affairs*, July/August 2015, www.foreignaffairs.com/articles/2015-06-16/will-humans-go-way-horses (accessed 21 February 2019); This quote has often been attributed to Mandela, but the evidence for this is unclear; Alan Kay, 'Learning vs. Teaching with Educational Technologies', *EDUCOM Bulletin*, 3/4 (1983), p.17.

Introduction

1 For example, far from seeing life expectancy surge to 120, in the USA life expectancy has actually been declining for the first time since the 1960s. Similarly, space and moon travel for humans (let alone serious off-planet manufacturing) remain very challenging and costly. Google's Lunar X Prize, which offered a prize for privately funded teams to land a robot on the moon, shifted its deadline back from December 2012 numerous times, and in March 2018 Google withdrew their US$30 million in prize funding altogether. Some companies have continued regardless, although the most advanced of these failed in its attempt to land a robot on the moon in April 2019. It is clear that space and moon travel for humans is likely to be only for the very wealthy few for quite some time.

2 Author interview with David Brougham, 2 May 2018.

3 H.A. Haenssle, C. Fink, R. Schneiderbauer, F. Toberer, T. Buhl, A. Blum, A. Kalloo, A. Ben Hadj Hassen, L. Thomas, A. Enk and L. Uhlmann, 'Reader study level-I and level-II Groups; Man against machine: Diagnostic performance of a deep learning convolutional neural network for dermoscopic melanoma recognition in comparison to 58 dermatologists', *Annals of Oncology*, 29, 8 (1 August 2018), pp.1836–42.

4 'The Automation Advantage', Alpha Beta, 2017, www.alphabeta.com/wp-content/uploads/2017/08/The-Automation-Advantage.pdf (accessed 17 December 2018).

5 This book focuses primarily on labour productivity, how much economic output we produce on average per hour. Multi-factor productivity is a closely related but slightly different measure that looks at how effectively we use our inputs of labour and capital to create output. It is typically calculated as the unexplained residual in growth accounting, the part of output that cannot be clearly attributed to workers' wages or capital investment, and is therefore sometimes referred to as a measure of our ignorance.

6 See Glenn Greenwald's TED talk for an excellent discussion of the dangers: 'Why privacy matters', TED, October 2014, www.ted.com/talks/

glenn_greenwald_why_privacy_matters (accessed 17 December 2018). On the relations between big data, psychological profiling, targeted marketing and capitalism see Shoshana Zuboff, *The Age of Surveillance Capitalism: The Fight for a Human Future at the New Frontier of Power,* Public Affairs, New York, 2019.

7 'Labour market statistics: September 2018 quarter', Statistics New Zealand, 6 November 2018, www.stats.govt.nz/informationreleases/labour-market-statisticsseptember-2018-quarter (accessed 15 December 2018).

8 Robert D. Atkinson, 'In defense of robots', *National Review*, 21 April 2017, www.nationalreview.com/2017/04/robots-jobs-industrial-future (accessed 17 December 2018). Wadhwa is not the only one making such bold predictions. Kai-Fu Lee, a former Apple and Google executive, told the CBS TV programme *60 Minutes* that 40 per cent of the jobs in the world would be displaced by technology within fifteen years. See 'Facial and Emotional Recognition: How One Man Is Advancing Artificial Intelligence', *60 Minutes*, 13 January 2019, www.cbsnews.com/news/60-minutes-ai-facial-and-emotional-recognition-how-one-man-is-advancing-artificial-intelligence (accessed 26 January 2019).

9 Alexandra Suich Bass, 'Non-tech businesses are beginning to use artificial intelligence at scale', *The Economist*, 31 March 2018, www.economist.com/specialreport/2018/03/31/non-techbusinesses-are-beginning-to-useartificial-intelligence-at-scale (accessed 17 December 2018).

10 Angus Deaton, *The Great Escape: Health, Wealth, and the Origins of Inequality*, Princeton University Press, Princeton, 2013. Charles Kenny has also made the same point in his book *Getting Better: Why Global Development Is Succeeding – And How We Can Improve the World Even More*, Basic Books, New York, 2012; and Steven Pinker has done the same in his book *Enlightenment Now: The Case for Reason, Science, Humanism, and Progress*, Penguin Books, New York, 2018.

11 For an excellent introduction to the issues in a New Zealand context see David Brougham, Jarrod Haar and Yumiko Oliver-Gray, 'Will my job be taken by a robot?', in *The Big Questions: What is New Zealand's Future?*, Penguin Random House, Auckland, 2018.

12 Suich Bass, 'Non-tech businesses are beginning to use artificial intelligence at scale'.

13 The former head of AI at Uber and professor at New York University Gary Marcus provides a critical overview of the limits of deep learning techniques in 'Deep Learning: A Critical Appraisal', ArXiv.org, https://arxiv.org/ftp/arxiv/papers/1801/1801.00631.pdf (accessed 12 December 2018).

Chapter One

1 See, for instance, 'Rise of the machines: How soon until robots take your job?', *NZ Herald*, 12 March 2018, www.nzherald.co.nz/business/news/article.cfm?c_id=3&objectid=12011383; Jamie Morton, 'The robots are coming – is your job safe?', *NZ Herald*, 25 March 2017, www.nzherald.co.nz/business/news/article.cfm?c_id=3&objectid=11823618; Henry Bodkin, 'Rise of the machines: robots set to take over in bed', Stuff, 20 December 2016, www.stuff.co.nz/life-style/love-sex/87791495/rise-of-the-machines-robots-set-to-take-over-in-bed; Dana McCauley, 'Robots threaten "millions" of jobs', *NZ Herald,* 3 May 2017, www.nzherald.co.nz/business/news/article.cfm?c_id=3&objectid=11849335; 'The robot that wants to start a family', *NZ Herald*, 25 November 2017, www.nzherald.co.nz/business/news/article.cfm?c_id=3&objectid=11947834

(all accessed 15 December 2018).

2 Robert Solow was awarded the Sveriges Riksbank Prize in Economic Sciences in Memory of Alfred Nobel (often known as the Nobel Memorial Prize in Economics) in 1987 but Trevor Swan's contributions were not recognised in the same way.

3 One feature of economics is the tendency to use jargon when plain language can sometimes explain the same thing. This has the unfortunate effect of locking people out of the ideas being discussed. Ha-Joon Chang, an economist at the University of Cambridge, is a prominent critic of this. He argues, and I agree, that these ideas are important and should be widely discussed even without specialist training. For further details see Ha-Joon Chang, *Economics: The User's Guide,* Bloomsbury Press, New York, 2014.

4 Ian Nichol, *Stand and Deliver: 40 Simple Steps to Successful Public Speaking,* Matador, Kibworth, 2018, p.251.

5 'They won this year's Nobel for economics. Here's why their work matters', World Economic Forum, www.weforum.org/agenda/2018/10/meet-the-winners-of-this-years-nobel-prize-for-economics (accessed 2 December 2018).

6 Christensen is best known for his theories of disruptive innovation, which look at how businesses enter new markets and sectors, but recently he has focused on how innovations differ when seen from the perspective of outcomes like jobs and growth.

7 This section draws on Clayton M. Christensen and Derek van Bever, 'The Capitalist's Dilemma', *Harvard Business Review,* 92, 6 (June 2014), pp.60–68, https://hbr.org/2014/06/the-capitalists-dilemma (accessed 2 December 2018). For further detail see Clayton M. Christensen, Efosa Ojomo and Karen Dillon, *The Prosperity Paradox: How Innovation Can Lift Nations Out of Poverty,* HarperBusiness, New York, 2019.

8 For other examples of market-creating innovation's impact on jobs see Bryan C. Mezue, Clayton M. Christensen and Derek van Bever, 'The Power of Market Creation', *Foreign Affairs,* 94, 1, January/February 2015, pp.69–76, https://www.foreignaffairs.com/articles/africa/2014-12-15/power-market-creation (accessed 2 December 2018).

9 Christensen and van Bever, 'The Capitalist's Dilemma'.

10 James Manyika, Susan Lund, Michael Chui, Jacques Bughin, Jonathan Woetzel, Parul Batra, Ryan Ko and Saurabh Sanghv, *Jobs Lost, Jobs Gained: Workforce Transitions in a Time of Automation,* McKinsey Global Institute, 2017, p.4, https://www.mckinsey.com/~/media/mckinsey/featured%20insights/future%20of%20organizations/what%20the%20future%20of%20work%20will%20mean%20for%20jobs%20skills%20and%20wages/mgi-jobs-lost-jobs-gained-report-december-6-2017.ashx (accessed 30 January 2019).

11 Market-creating innovation can be usefully seen as a way to generate 'new tasks' for workers to do, which Daron Acemoglu and his co-author Pascual Restrepo emphasise is critical to keeping employment high in the face of automation. Acemoglu has discussed this similarity in an interview with the American Enterprise Institute. For the original paper see Daron Acemoglu and Pascual Restrepo, 'The Race between Man and Machine: Implications of Technology for Growth, Factor Shares, and Employment', *American Economic Review,* 108, 6 (2018), pp.1488–542. For a discussion of the connections with Christensen's terminology of market-creating innovation see James Pethokoukis, 'Will technology enable workers or replace them? A long-read Q&A with Daron Acemoglu', American Enterprise Institute Ideas, www.aei.org/publication/will-technology-

enable-workers-or-replace-them-a-long-read-qa-with-daron-acemoglu (accessed 2 December 2018).

12 Joseph Schumpeter, the father of much modern thinking about innovation, emphasised the differences between innovation by incumbent firms and innovation by new entrants through creative destruction. Another common distinction is between product and process innovation.

13 Pethokoukis, 'Will technology enable workers or replace them?'

14 Marco Vivarelli, 'Innovation, Employment, and Skills in Advanced and Developing Countries: A Survey of Economic Literature', *Journal of Economic Issues,* 48, 1 (2014), pp.123–54.

15 Matthias Buerger, Tom Broekel and Alex Coad, 'Regional dynamics of innovation: Investigating the coevolution of patents, research and development (R&D), and employment', *Regional Studies,* 46, 5 (2012), pp.565–82. See also Francesco Bogliacino, Mariacristina Piva and Marco Vivarelli, 'R&D and employment: An application of the LSDVC estimator using European microdata', *Economics Letters,* 116, 1 (2012), pp.56–59.

16 Marco Vivarelli, 'Innovation and employment', *IZA World of Labor,* 154 (2015), https://wol.iza.org/uploads/articles/154/pdfs/innovation-and-employment.pdf (accessed 2 December 2018).

17 Kevin Jenkins, 'Platforms in Aotearoa: Our fast growing sharing economy', *Policy Quarterly,* 14, 1 (2018), pp.10–17.

18 Christensen himself has argued that Uber is a sustaining innovation; see Clayton M. Christensen, Michael E. Raynor and Rory McDonald, 'What Is Disruptive Innovation?', *Harvard Business Review,* 93, 12 (December 2015), pp.44–53.

19 Shannon Bond, 'Fare cutting puts the brakes on Uber's revenue growth', *Financial Times,* 15 February 2019, https://www.ft.com/content/aa76c512-30eb-11e9-8744-e7016697f225 (accessed 27 March 2019).

20 See Leslie Hook, 'Can Uber ever make money?', *Financial Times,* 23 June 2017, www.ft.com/content/09278d4e-579a-11e7-80b6-9bfa4c1f83d2 (accessed 2 December 2018).

21 'Firms that burn up $1bn a year are sexy but statistically doomed', *The Economist,* 21 October 2017, www.economist.com/news/business/21730446-five-outlierschesapeake-energy-netflix-nexteraenergy-tesla-and-uber-have-collectively (accessed 2 December 2018).

22 In fact, more broadly, there is evidence that product innovation favours the wealthy, who, after all, represent a more profitable market. See Xavier Jaravel, 'The Unequal Gains from Product Innovations: Evidence from the U.S. Retail Sector', *Quarterly Journal of Economics,* October 2018, https://doi.org/10.1093/qje/qjy031 (accessed 23 January 2019).

23 For malaria see 'Malaria's Impact Worldwide', Centers for Disease Control and Prevention, www.cdc.gov/malaria/malaria_worldwide/impact.html (accessed 3 December 2018). For AIDS see 'HIV/AIDS', Centers for Disease Control and Prevention, www.cdc.gov/hiv/statistics/overview (accessed 3 December 2018).

24 Michael Kremer and Rachel Glennerster, *Strong Medicine: Creating Incentives for Pharmaceutical Research on Neglected Diseases,* Princeton University Press, Princeton, New Jersey, 2004, pp.24–29.

25 Most Africans afflicted with river blindness simply would not have been able to pay for the drugs otherwise and Merck's efforts to find an external public or non-profit funder were unsuccessful. See Yolanda Hernando, Kaela Cowell and Brian D. Wright, 'Doing well while fighting

river blindness: The alignment of a corporate drug donation programme with responsibilities to shareholders', *Tropical Medicine & International Health*, 21 (2016), pp.1304–10, https://onlinelibrary.wiley.com/doi/full/10.1111/tmi.12759 (accessed 3 December 2018).

26 Ibid., p.1308.

27 For a useful introduction to the literature on the role of profits as a motive in technological change see Petra Moser, 'How Do Patent Laws Influence Innovation? Evidence from Nineteenth Century World's Fairs', *American Economic Review*, 95, 4 (2005), pp.1214–36; see also Kenneth Sokoloff, 'Inventive Activity in Early Industrial America: Evidence from Patent Records, 1790–1846', *Journal of Economic* History, 48, 4 (1988), pp.813–50.

28 See Acemoglu and Restrepo, 'The Race between Man and Machine', for a discussion of how high wages can incentivise automation. Note that Daron Acemoglu's work has previously emphasised both factor prices and market size for potential new innovations as a determinant of the direction of technical change but his recent work with Pascual Restrepo looking at the determinants of automation emphasises factor prices as particularly important. He also confirmed this emphasis in an interview with the author.

29 This is for large cafés and restaurants; see 'Business Performance Benchmarker Cafes and Restaurants', Statistics New Zealand, https://shinyapps.stats.govt.nz/bpb (accessed 3 December 2018).

30 Author interview with Daron Acemoglu, 24 April 2018.

31 Joseph Stiglitz, 'The coming great transformation', *Journal of Policy Modelling*, 39, 4 (2017), p.631.

32 As firms decide how to invest to reduce their production costs it is the relative cost of each input that matters. If carbon is priced too low they may invest in technologies for labour cost reductions rather than carbon reductions. The same logic applies for investment decisions to fund the invention and development of technologies that could save on different inputs.

33 Daron Acemoglu and Pascual Restrepo, 'Artificial Intelligence, Automation and Work', in Ajay Agarwal, Avi Goldfarb and Joshua Gans (eds), *Economics of Artificial Intelligence: An Agenda*, University of Chicago Press, Chicago, forthcoming, p.15, www.nber.org/chapters/c14027.pdf (accessed 29 January 2019).

34 On this point I have benefited from discussions with Lant Pritchett of the University of Oxford.

35 NBER, *The Rate and Direction of Inventive Activity: Economic and Social Factors*, Princeton University Press, Princeton, New Jersey, 1962.

36 Anthony Atkinson, *Inequality – What Can Be Done?*, Harvard University Press, Cambridge, Massachusetts, 2015, p.30.

37 Daron Acemoglu, Philippe Aghion, Leonardo Bursztyn and David Hemous, 'The Environment and Directed Technical Change', *American Economic Review*, 102, 1 (2012), pp.131–66.

38 Author interview with Phil O'Reilly, 10 May 2018.

39 Ricardo Hausmann and Dani Rodrik, *Doomed to Choose: Industrial Policy as Predicament*, John F. Kennedy School of Government, Harvard University, Cambridge, Massachusetts, 2006, https://drodrik.scholar.harvard.edu/files/dani-rodrik/files/doomed-to-choose.pdf (accessed 3 December 2018).

40 Stiglitz, 'The coming great transformation', p.636.

41 Richard Newell, Adam B. Jaffe and Robert N. Stavins, 'The Induced Innovation Hypothesis and Energy-Saving Technological Change', *Quarterly Journal of Economics*, 114, 3 (1999), pp.941–75.

42 David Popp, 'Induced Innovation and Energy Prices', *American Economic Review*, 92, 1 (2002), pp.160–80. He also finds that the quality of existing knowledge has a strong influence on innovation.

43 Acemoglu et al., 'The Environment and Directed Technical Change', p.132.

44 Zorina Khan and Kenneth L. Sokoloff, 'Institutions and Democratic Invention in 19th-Century America: Evidence from "Great Inventors," 1790–1930', *American Economic Review*, 94, 2 (2004), pp.395–401.

45 For an overview of the critical view see Joseph E. Stiglitz, 'Economic Foundations of Intellectual Property Rights', *Duke Law Journal*, 57, 6 (2008), pp.1693–724.

46 Others have argued that approaches like compulsory licensing, which forces patent owners to license the new inventions to other firms, including potential competitors, might be more effective at encouraging innovation. See Petra Moser, 'Patents and Innovation in Economic History', *Annual Review of Economics*, 8, 1 (2016), pp.241–58.

47 This is partly because income tax is easier to collect and less subject to flight (although many countries tax capital more highly than we do), but also because of policy choices on capital gains and other corporate taxes. There are other considerations for how and why we tax but if we are concerned about the kind of innovation we may be heading for, we may need to think again about our tax system.

48 Kate Raworth, *Doughnut Economics: Seven Ways to Think Like a 21st Century Economist*, Chelsea Green Publishing, Vermont, 2017, loc. 3587 of 4507, Kindle.

49 The double-declining-balance method is applicable to most plant and equipment and allows businesses to get the tax deduction quickly. In addition, firms can deduct interest payments to debt-holders from their taxable earnings. Given debt is largely used to finance capital investments not hire workers, this tilts in favour of capital further. For a critique of this see 'A senseless subsidy', *The Economist*, 16 May 2015, www.economist.com/briefing/2015/05/16/a-senseless-subsidy (accessed 16 December 2018). While there was no capital gains tax at the time of writing, the extended bright line test means profits on investment properties sold within five years of purchase can be taxed.

50 Acemoglu and Restrepo, 'Artificial Intelligence, Automation and Work'.

51 To be clear, New Zealand has non-residential business investment that is slightly below the OECD average. Yet that may in part reflect that most countries tilt the playing field even further toward capital than we do. For example, these distortions are even more significant in the USA. Daron Acemoglu explains, 'We implicitly subsidize production with machines relative to production with labor. If you buy a machine, you don't have to pay payroll taxes, you can debt finance it, and that's going to get a subsidy from the government. And the capital income that is generated is going to be taxed at a lower rate. If you hire a worker, you have to pay payroll taxes, they are going to be taxed at a much higher rate, you have to put up with lots of other costs coming from regulations, and other things you have to do when you're employing workers. So that creates a very non-level playing field, so we're essentially subsidizing firms implicitly for using machines rather than labor. So I think it's probably a good idea to start thinking about how we can even that playing field a little bit.' See James Pethokoukis, 'On automation and the future of work: A short-read Q&A with economist Daron Acemoglu', American Enterprise Institute, www.aei.org/publication/on-automation-and-the-

future-of-work-a-short-read-qa-with-economist-daron-acemoglu (accessed 16 December 2018).

52 Dani Rodrik, 'The Good Jobs Challenge', Project Syndicate, www.project-syndicate.org/commentary/how-countries-can-create-middle-class-jobs-by-dani-rodrik-2019-02 (accessed 21 February 2019).

53 Labour-market imperfections and frictions imply that the market-clearing wage (the wage that occurs in a competitive market) is above the social opportunity cost of labour. Finding a job is a complicated and slow process, so for people to work they have to be paid more than what they actually value any given hour of their own time at (the opportunity cost of their labour). This higher wage is to compensate for the costs of searching for a job. Firms decide whether to automate a given task based on this actual wage cost, but a utilitarian social planner would use a lower hypothetical wage (economists call this a shadow wage), which instead reflects the opportunity cost of an hour of workers' time. Businesses therefore may choose to automate when it is not socially optimal. See Acemoglu and Restrepo, 'Artificial Intelligence, Automation and Work'.

54 Katie Roof, 'Elon Musk says "humans are underrated," calls Tesla's "excessive automation" a "mistake"', Tech Crunch, https://techcrunch.com/2018/04/13/elon-musk-says-humans-are-underrated-calls-teslas-excessive-automation-a-mistake (accessed 4 December 2018).

55 Easy monetary policy could also have made it easier even for firms that automate excessively to stay in business by allowing them to sustain themselves through cheap capital – meaning it is possible to have the worst of both worlds: fewer jobs but less output and thus less productivity growth.

56 See, for example, Nick Srnicek, *Inventing the Future: Postcapitalism and a World Without Work*, Verso, London, 2016; 'The Automation Advantage', AlphaBeta, 2017, www.alphabeta.com/wp-content/uploads/2017/08/The-Automation-Advantage.pdf (accessed 15 December 2018); Kevin J. Delaney, 'The robot that takes your job should pay taxes, says Bill Gates', *Quartz*, 17 February 2017, https://qz.com/911968/bill-gates-the-robot-that-takes-your-job-should-pay-taxes (accessed 15 December 2018).

57 For a detailed discussion of the state's role in supporting the technology used in the iPhone see Mariana Mazzucato, 'The State Behind the iPhone', in *The Entrepreneurial State: Debunking Public vs. Private Sector Myths*, Anthem Press, London, 2013, ch.5.

58 'Research and development in New Zealand: 2014', Stats NZ, http://archive.stats.govt.nz/browse_for_stats/businesses/research_and_development/research-development-nz-2014/international-r-and-d-activity.aspx (accessed 15 December 2018).

59 Christensen and van Bever, 'The Capitalist's Dilemma'.

60 Mariana Mazzucato, *The Entrepreneurial State: Debunking Public vs. Private Sector Myths*, Anthem Press, London, 2013, loc. 496 of 3988, Kindle.

61 Mariana Mazzucato, *Mission-Oriented Research & Innovation in the European Union*, European Commission, Brussels, 2018, p.4, https://ec.europa.eu/info/sites/info/files/mazzucato_report_2018.pdf (accessed 18 December 2018).

62 Kerstine Appunn and Julian Wettengel, 'Germany's greenhouse gas emissions and climate targets', *Clean Energy Wire*, 9 July 2018, www.cleanenergywire.org/factsheets/germanys-greenhouse-gas-emissions-and-climate-targets (accessed 15 December 2018).

63 'Global Innovation Index 2018',

Global Innovation Index, https://www.globalinnovationindex.org/gii-2018-report# (accessed 18 December 2018).

64 'Gross domestic spending on R&D (indicator)', OECD, 2018, https://doi.org/10.1787/d8b068b4-en (accessed 16 December 2018).

65 Charlie Dreaver, 'Inspire more women and more Māori and Pasifika into science', Radio New Zealand, 1 October 2018, www.radionz.co.nz/news/political/365857/inspire-more-women-and-more-maori-and-pasifika-into-science (accessed 15 December 2018).

66 The Energy and Minerals Research Fund has now been folded into the Endeavour Fund.

67 *National Statement of Science Investment 2015–2025*, Ministry of Business, Innovation and Employment, Wellington, 2015, p.34.

68 Mezue, Christensen and van Bever, 'The Power of Market Creation'.

69 See 'Direction to Callaghan Innovation – Criteria for Assessing Proposals for Funding Project and Student Grant Funding', gazette.govt.nz, https://gazette.govt.nz/notice/id/2018-go4864 (accessed 3 February 2019). These Callaghan R&D Project Grants will continue; however, Callaghan's larger R&D Growth Grant programme is being replaced by an R&D tax credit. Neither the criteria for R&D Growth Grants that are now being phased out, nor the R&D tax credit, put any premium on the job-creating impact of the innovations they fund with public money.

70 See 'Endeavour Fund 2019 Funding Round', gazette.govt.nz, https://gazette.govt.nz/notice/id/2018-go4196 (accessed 3 February 2019). This is not to say that all science funding should consider employment, but the Endeavour Fund is specifically aimed at research that has high potential to benefit and positively transform New Zealand.

71 As Jane Halteh and colleagues have written, '[T]here are also concerns that a small domestic market and insufficient competition within New Zealand will generate insufficient innovation and growth to compensate for technological job losses.' This underlines the need for New Zealand in particular to push our science and innovation system toward market-creating innovation to create new jobs rather than just leaving it up to the market. See Jane Halteh, Jim Arrowsmith, Jane Parker, Theodore E. Zorn and Tim Bentley, 'The impact of technology on employment: a research agenda for New Zealand and beyond', *Labour & Industry: A Journal of the Social and Economic Relations of Work*, 28, 3 (2018), p.206.

72 'National Science Challenges', Ministry of Business, Innovation and Employment, www.mbie.govt.nz/science-and-technology/science-and-innovation/funding-information-and-opportunities/investment-funds/national-science-challenges (accessed 18 December 2018).

73 Quoted in Ralph Gomory and Richard Sylla, 'The American Corporation', *Daedalus*, 142, 2 (2013), p.107.

74 Ibid., p.110.

75 Judy Brown, 'Social Responsibility and the Company: A New Perspective on Governance, Strategy and the Community', *Social Policy Journal of New Zealand*, 9 (November 1997).

76 See Raworth, *Doughnut Economics*, loc. 2861 of 4507, Kindle.

77 A focus on short-term gain over long-term planning and goals.

78 For an inside story of how short-termism and the pressure to maximise shareholder value pushes towards greater automation, see Kevin Roose, 'The Hidden Automation Agenda of the Davos Elite', *New York Times*, 25 January 2019, www.nytimes.com/2019/01/25/technology/automation-davos-world-economic-forum.html (accessed 26 January 2019).

79 Andrew Haldane and Richard Davies,

'The Short Long', paper presented at the 29th SUERF Colloquium, Brussels, 11 May 2011, www.bis.org/review/r110511e.pdf (accessed 18 December 2018).

80 Christensen and van Bever, 'The Capitalist's Dilemma'; for a discussion of the broader problems this generates, see Lynn Stout, *The Shareholder Value Myth: How Putting Shareholders First Harms Investors, Corporations, and the Public*, Berrett-Koehler Publishers, San Francisco, 2012.

81 There are other reasons why executives may choose buy-backs, but, as William Lazonick lays out, they are hardly convincing. William Lazonick, 'Profits Without Prosperity', *Harvard Business Review*, September 2014, https://hbr.org/2014/09/profits-without-prosperity (accessed 17 December 2018).

82 'Record share buybacks should be raising alarms', *Financial Times*, 7 November 2018, www.ft.com/content/11fd4cb6-e1db-11e8-8e70-5e22a430c1ad (accessed 17 December 2018). Another study from the Roosevelt Institute found that between 2004 and 2014 publicly listed American companies spent on average three-fifths of their profits on share buy-backs, and since then buy-backs have increased. A recent study by the US Federal Reserve showed that most of the money firms repatriated after the tax breaks provided by the Trump administration have been spent on share buy-backs rather than on R&D or growing the business; see Noel Randewich, 'Little evidence of capex surge from tax cuts: Fed', Reuters, 4 September 2018, www.reuters.com/article/us-usa-stocks-buybacks/little-evidence-of-capex-surge-from-tax-cuts-fed-idUSKCN1LK2JD (accessed 17 December 2018).

83 Brian Gaynor, 'Share buybacks can do more harm than good', *NZ Herald*, 30 April 2016, www.nzherald.co.nz/business/news/article.cfm?c_id=3&objectid=11631082 (accessed 17 December 2018); Chalkie, 'Value of buybacks tough to figure out', Stuff, 27 November 2013, www.stuff.co.nz/business/opinion-analysis/9445504/Value-of-buybacks-tough-to-figure-out (accessed 17 December 2018).

84 Lazonick, 'Profits Without Prosperity'. Lazonick highlights how this is the case in alternative energy, nanotechnology and pharmaceutical companies.

85 Katy Milani and Irene Tung, *Curbing Stock Buybacks: A Crucial Step to Raising Worker Pay and Reducing Inequality*, Roosevelt Institute, July 2018, http://rooseveltinstitute.org/wp-content/uploads/2018/07/The-Big-Tradeoff-Report_072618.pdf (accessed 27 March 2019).

86 Dominic Barton, James Manyika, Timothy Koller, Robert Palter, Jonathan Godsall and Joshua Zoffer, 'Measuring the Economic Impact of Short-termism', McKinsey Global Institute discussion paper, February 2017, www.mckinsey.com/~/media/mckinsey/featured%20insights/long%20term%20capitalism/where%20companies%20with%20a%20long%20term%20view%20outperform%20their%20peers/mgi-measuring-the-economic-impact-of-short-termism.ashx (accessed 17 December 2018).

87 They use the rolling five-year performance of the S&P 500 as their benchmark.

88 Chris Flood, 'NZ Super is the All Blacks of the sovereign wealth fund world', *Financial Times*, 8 October 2017, https://www.ft.com/content/871bd0dc-a828-11e7-93c5-648314d2c72c (accessed 13 March 2019).

89 This long-term focus could be strengthened further by the big beasts of investment in turn measuring companies against metrics that are inherently longer-term. Example metrics might include how many patents a company has registered,

how energy-efficient their production is, and their ten-year economic value added, which measures how and where companies have truly created wealth. See Dominic Barton and Mark Wiseman, 'Focusing Capital on the Long Term', *Harvard Business Review*, January–February 2014, https://hbr.org/2014/01/focusing-capital-on-the-long-term (accessed 17 December 2018).

90 Eric Beinhocker and Nick Hanauer, 'Redefining Capitalism', *McKinsey Quarterly*, September 2014, www.mckinsey.com/global-themes/long-term-capitalism/redefining-capitalism (accessed 17 December 2018).

91 Author interview with Diane Ruwhiu, 12 December 2018.

92 According to the 2018 Acumen Edelman Trust Barometer, 'The Battle for Truth: 2018 Acumen Edelman Trust Barometer', Acumen Republic, March 2018, www.acumenrepublic.com/media/1406/trust-barometer-new-zealand-march-2018.pdf (accessed 17 December 2018).

93 2018 Acumen Edelman Trust Barometer, '2018 Edelman Trust Barometer: The State of Trust in Business', Edelman, January 2018, www.edelman.com/sites/g/files/aatuss191/files/2018-10/2018_Edelman_Trust_Barometer_State_of_Business.pdf (accessed 19 January 2019).

94 *Financial Times*, 'Knowledge is the New Capital: Why companies keep making one simple and costly mistake', YouTube, 26 June 2017, www.youtube.com/watch?v=lu8TXmOVHGg&feature=youtu.be (accessed 17 December 2018).

95 For the full scope of her work see Zeynep Ton, *The Good Jobs Strategy: How the Smartest Companies Invest in Employees to Lower Costs and Boost Profits*, Houghton Mifflin Harcourt, Boston, 2014.

96 Bill Cochrane, Michael Fletcher, Gail Pacheco and Alexander Plum, 'Low pay in NZ', Ministry of Business, Innovation and Employment, January 2018, www.mbie.govt.nz/assets/c92012e10c/low-pay-in-nz-january-2018.pdf (accessed 26 January 2019).

97 This problem is made more acute by the rise of high-frequency trading.

98 Dominic Barton, 'Capitalism for the Long Term', *Harvard Business Review*, March 2011, https://hbr.org/2011/03/capitalism-for-the-long-term (accessed 27 March 2019).

99 For a discussion of how the creation of new tasks and jobs can increase multi-factor productivity by allowing labour to focus more on tasks in which it has a comparative advantage (thereby improving the efficiency of use of labour and capital in production), see Daron Acemoglu and Pascual Restrepo, 'Automation and New Tasks: How Technology Displaces and Reinstates Labor', NBER Working Paper 25684, March 2019, https://www.nber.org/papers/w25684 (accessed 4 April 2019).

Chapter Two

1 Siemens Werner, 'On the Electromotive Action of Illuminated Selenium, Discovered by Mr Fritts of New York', *English Mechanics and the World of Science*, 41, 1,956 (1885), p.340, https://books.google.co.nz/books?id=PAPI9Ej0VncC&pg=PA340&lpg=PA340&dq#v=onepage&q&f=false (accessed 4 December 2018).

2 Both quotes are reported in John Perlin, *Let It Shine: The 6,000 Year Story of Solar Energy*, New World Library, Novato, California, 2013, p.314.

3 'Share of electricity generation by fuel', International Energy Agency, www.iea.org/statistics/?country=WORLD&year=2016&category=Electricity&indicator=undefined&mode=chart&dataTable=ELECTRICITYANDHEAT (accessed 6 February 2019).

4 Manchester City, for everyone who is not a football fan like me.

5 Another example is 3D printing.

The first 3D printer was invented by Chuck Hall in 1983, well before I was born. By 2018, thirty-five years later, the impact of 3D printing is growing strongly, but it is still far easier to read about how we are on the verge of a 3D-printing boom than to see its large-scale applications. See Bryan Borzykowski, 'Why 3D is more than a passing fad for jobs', BBC, 28 January 2016, www.bbc.com/capital/story/20160127-model-your-career-in-3-d (accessed 15 December 2018); and Rakesh Sharma, 'What Happened to the 3D Printing Revolution? (SSYS, XONE)', Investopedia, 17 August 2016, www.investopedia.com/articles/investing/081715/what-happened-3d-printing-revolution.asp (accessed 15 December 2018). There have also been some substantial stock crashes for 3D printing companies. Kevin Kelleher, 'Was 3D Printing Just a Passing Fad?', *Time*, 10 June 2015, http://time.com/3916323/3d-printer-stocks (accessed 15 December 2018). The impact may eventually arrive, but it is taking time.

6 Cited in Susan Ratcliffe (ed.), *Oxford Essential Quotations*, 4th edn, Oxford University Press, 2016, www.oxfordreference.com/view/10.1093/acref/9780191826719.001.0001/q-oro-ed4-00010666 (accessed 6 February 2019).

7 Janosch Delcker, 'The man who invented the self-driving car (in 1986)', Politico, 19 July 2018, www.politico.eu/article/delf-driving-car-born-1986-ernst-dickmanns-mercedes (accessed 3 February 2019).

8 The working paper came out in 2013 but the final published version is Carl Benedikt Frey and Michael A. Osborne, 'The future of employment: How susceptible are jobs to computerisation?', *Technological Forecasting and Social Change*, 114 (2017), pp.254–80.

9 Laurence Kubiak, Mark Dean and Aaron Drew, *Disruptive Technologies Risks, Opportunities – Can New Zealand make the most of them?*, NZIER, 21 October 2015, https://nzier.org.nz/static/media/filer_public/6d/6e/6d6ecf8b-032c-4551-b0a7-8cd0f39e2004/disruptive_technologies_for_caanz.pdf (accessed 15 December 2018).

10 I was at a conference on the future of work where a leading US academic observed saltily that it took Frey and Osborne four years to find a journal to publish their working paper, which was not peer-reviewed, and when they did it was a little-known journal. This, he mused, well sums up the methodological shortcomings.

11 Frey and Osborne surveyed experts about the automatability of 70 occupations only, and then used machine learning techniques to impute the results for the remaining 632 occupations they consider in the study. This added another potential source of overestimation.

12 There is usually a caveat to be found somewhere in the studies about the commercial viability and possible speed of adoption of new technologies, but not in the opening paragraphs. These caveats tend to disappear completely from the secondary coverage of the reports in popular media.

13 It is also worth noting, as David Autor does, the 'long history of leading thinkers overestimating the potential of new technologies to substitute for human labor and underestimating their potential to complement it'. See David Autor, 'Polanyi's Paradox and the Shape of Employment Growth', NBER Working Paper 20,485, 2014, p.32, www.nber.org/papers/w20485.pdf (accessed 16 December 2018).

14 See for example Kubiak et al., *Disruptive Technologies Risks, Opportunities*; Gareth Kiernan, *From Education to Employment: Megatrends Affecting NZ's Working Environment*, Infometrics, 2017, http://static.infometrics.co.nz/Content/Megatrends_web%20version.pdf (accessed 6 February 2019). Some other studies acknowledge much

more clearly that they are only talking about technological feasibility, not likely adoption, but these warnings are not often included in the main findings or press releases, and the studies do not attempt to go beyond an acknowledgement to actually model adoption speeds. See for example Mark Muro, Robert Maxim and Jacob Whiton, *Automation and Artificial Intelligence: How Machines Are Affecting People and Places*, Brookings Institution, January 2019, www.brookings.edu/wp-content/uploads/2019/01/2019.01_BrookingsMetro_Automation-AI_Report_Muro-Maxim-Whiton-FINAL-version.pdf (accessed 24 January 2019). This states on page 20, 'We have no ability to assess how the gap between "automation potential" (which our figures are based on) and "actual adoption" (what will actually occur) will evolve, so our estimates should be seen as an upper bound – perhaps even an extreme one. Our figures describe what is possible – not what is likely.'

15 Author interview with Damon Acemoglu, 24 April 2018.

16 'Small Business in New Zealand', Ministry of Business, Innovation and Employment, June 2017, www.beehive.govt.nz/sites/default/files/2017-12/Small%20Business%20-%20Annex%203%20Small%20Business%20Factsheet.pdf (accessed 16 December 2018).

17 Jo Moir, 'Novopay botch-ups cost $45m to fix', Stuff, 18 February 2018, www.stuff.co.nz/national/education/66349800/novopay-botchups-cost-45m-to-fix (accessed 15 December 2018).

18 David Johnson, 'These Are the Most Productive Countries in the World', *Time*, 4 January 2017, http://time.com/4621185/worker-productivity-countries (accessed 15 December 2018). Cross-country productivity comparisons are particularly difficult; this is based on GDP per hour calculated from the OECD data. 'Level of GDP per capita and productivity', OECD.Stat, http://stats.oecd.org/Index.aspx?DataSetCode=PDB_LV (accessed 15 December 2018).

19 Geoff Mason, 'Investigating New Zealand-Australia productivity differences: New comparisons at industry level', New Zealand Productivity Commission Working Paper 2013/02, https://www.productivity.govt.nz/sites/default/files/NZPC-NZ-Australia-productivity-working-paper-2013-02.pdf (accessed 27 March 2019).

20 Ibid.

21 Notably, in comparing New Zealand and Australia these two factors explain most of the difference. Skills differences between the two countries explain very little variation in output.

22 Paul Conway, 'Achieving New Zealand's productivity potential', New Zealand Productivity Commission Research Paper 2-16/1, Wellington, 2016, p.22.

23 OECD, *The Future of Productivity*, OECD Publishing, Paris, 2015.

24 Conway, 'Achieving New Zealand's productivity potential', p.23.

25 'International Comparisons of Hourly Compensation Costs in Manufacturing, 2016 – Summary Tables', The Conference Board, www.conference-board.org/ilcprogram/index.cfm?id=38269#Table1 (accessed 15 December 2018).

26 It is even less persuasive in explaining the gulf we see in technological uptake and productivity *within* countries where conditions are similar. Ricardo Hausmann notes that in Mexico the state of Guerrero has a GDP per capita of US$5,281 but the state of Nuevo Leon has a GDP per capita of US$42,281. This is a result of enormous productivity differences. Yet as he points out, this productivity difference exists despite the same legal framework, judicial system, political system, exchange rate,

interest rate, macroeconomic setup and financial architecture.

27 Author interview with Ricardo Hausmann, 23 April 2018.

28 Ibid.

29 Ibid.

30 Brian Palmer, 'How Did Detroit Become Motor City?', *Slate*, www.slate.com/articles/news_and_politics/explainer/2012/02/why_are_all_the_big_american_car_companies_based_in_michigan_.html (accessed 15 December 2018).

31 David Laws, 'Fairchild, Fairchildren, and the Family Tree of Silicon Valley', Computer History Museum, 20 December 2016, www.computerhistory.org/atchm/fairchild-and-the-fairchildren (accessed 15 December 2018).

32 Author interview with Hausmann.

33 Author interview with Diego Comin, 16 May 2018.

34 The following examples are drawn from the author's discussions with Ricardo Hausmann and from his presentations.

35 E. Hornung, 'Immigration and the diffusion of technology: The Huguenot diaspora in Prussia', *American Economic Review*, 104, 1 (2014), pp.84–122.

36 Petra Moser, Alessandra Voena and Fabian Waldinger, 'German Jewish Émigrés and US Invention', *American Economic Review*, 104, 10 (2014), pp.3222–55.

37 This study refers to Total Factor Productivity growth. See Giovanni Peri, Kevin Shih and Chad Sparber, 'STEM Workers, H-1B Visas, and Productivity in US Cities', *Journal of Labor Economics*, 33, S1 (2015), pp.S225–55.

38 Keith McLeod, Richard Fabling and David Maré, 'Hiring New Ideas: International Migration and Firm Innovation in New Zealand', Motu Working Paper 14-14, 2014, p.iii, http://dx.doi.org/10.2139/ssrn.2531479 (accessed 16 December 2018).

39 'Long Term Skill Shortage List', Immigration New Zealand, February 2018, http://skillshortages.immigration.govt.nz/long-term-skill-shortage-list.pdf (accessed 15 December 2018).

40 While some of these may bring interesting technologies with them, it is unlikely this is a rich source of cutting-edge technology for New Zealand. See 'Occupation and Region of Employment for approved Resident Visa principal applicants', Immigration New Zealand, www.immigration.govt.nz/documents/statistics/statistics-occupation-region-for-resident-principla-applicants (accessed 5 February 2019). The rise in the number of points required for skilled immigrants that was introduced in October 2016 (from 140 to 160) may be changing this somewhat, but appears to be reducing the total number of visas rather than substantially altering the composition of occupations. Between 2015/16 and 2017/18 total skilled resident visas fell from 13,088 to 8,402. Yet general retail manager, café or restaurant manager and chef were still in the top four (nurses were fourth in 2015/16 and then first in 2017/18); these three still collectively accounted for 11 per cent of total skilled migrant resident visas. For a more detailed discussion and evidence of how this has in fact been a long-standing pattern, see Michael Reddell, 'Who has been getting residence visas?', Croaking Cassandra, 29 June 2017, https://croakingcassandra.com/2017/06/29/who-has-been-getting-residence-visas (accessed 15 December 2018).

41 See profiles at 'Piripiri (Cohort 2)', Edmund Hillary Fellowship, https://stories.ehf.org/cohort-2/home (accessed 16 December 2018).

42 Niall McCarthy, 'The Countries with the Most Native-Born People Living Abroad [Infographic]', *Forbes*, 15 January 2016, www.forbes.com/sites/niallmccarthy/2016/01/15/

the-countries-with-the-most-native-born-people-living-abroad-infographic/#2dfa53de7924 (accessed 15 December 2018).

43 AnnaLee Saxenian, *The New Argonauts: Regional Advantage in a Global Economy*, Harvard University Press, Cambridge, Massachusetts, 2007.

44 Data from 'Global Flow of Tertiary-Level Students', UNESCO, http://uis.unesco.org/en/uis-student-flow (accessed 15 December 2018).

45 Almost half of New Zealand's overseas students go to Australia. Even China, where average income is half of New Zealand's, has about the same share of its tertiary students studying abroad, with the USA again the number one destination. Many European Union countries such as France, Germany, Finland, Austria and Switzerland also send more students abroad than New Zealand does. Many of these go to other European Union countries such as the United Kingdom, but equally New Zealand sends almost 50 per cent of its overseas students to Australia.

46 Ricardo Hausmann, 'Should Business Travel Be Obsolete?', Project Syndicate, 20 January 2016, https://growthlab.cid.harvard.edu/news/should-business-travel-be-obsolete (accessed 15 December 2018).

47 Conway, 'Achieving New Zealand's productivity potential', pp.27–29.

48 OECD, *OECD Economic Surveys: New Zealand 2017*, OECD Publishing, Paris, 2017, p.36.

49 Simon Thomsen, 'Watch: Domino's just delivered its first pizza by drone', *Business Insider Australia*, 16 November 2016, www.businessinsider.com.au/dominos-just-became-the-first-business-to-deliver-food-by-drone-2016-11 (accessed 16 December 2018).

50 Rob Maetzig, 'Electric vehicles: Can we become the Norway of the Southern Hemisphere?', Stuff, 27 September 2017, www.stuff.co.nz/motoring/news/97198599/electric-vehicles-can-we-become-the-norway-of-the-southern-hemisphere (accessed 15 December 2018).

51 Tim Higgins and Charles Rollet, 'Tesla Sales Fall to Zero in Hong Kong after Tax Break Is Slashed', *Wall Street Journal*, 9 July 2017, www.wsj.com/articles/teslas-hong-kong-sales-gutted-by-tax-change-1499598003 (accessed 15 December 2018).

52 E.F. Schumacher, *Small Is Beautiful: Economics as if People Mattered*, Blond & Briggs, London, 1973.

53 Paul Conway and G. Zheng, 'Trade Over Distance for New Zealand Firms: Measurement and implications', New Zealand Productivity Commission Working Paper 2014/5, Wellington, 2014, www.productivity.govt.nz/sites/default/files/nzpc-working-paper-2014-05-trade-over-distance-rev2.pdf (accessed 6 February 2019).

54 Lisa Meehan, 'Productivity in New Zealand: The role of resource allocation among firms', *New Zealand Economic Papers*, 2019, https://www.tandfonline.com/doi/full/10.1080/00779954.2019.1573846 (accessed 10 March 2019).

55 Paul Conway, 'Can the Kiwi Fly? Achieving Productivity Lift-off in New Zealand', *International Productivity Monitor*, 34 (Spring 2018), p.59.

56 Hamish Rutherford, 'Adrian Orr says Super Fund "gave up" trying to invest in Christchurch rebuild', Stuff, 15 May 2018, www.stuff.co.nz/business/103902717/adrian-orr-says-super-fund-gave-up-trying-to-invest-in-christchurch-rebuild (accessed 15 December 2018).

57 Conway, 'Achieving New Zealand's productivity potential', p.37. For a more detailed discussion, see Geoff Mason and Matthew Osborne, 'Productivity, capital-intensity and labour quality at sector level in New Zealand and the UK', New Zealand Treasury Working Paper

07/01, Wellington, 2007, https://treasury.govt.nz/publications/wp/productivity-capital-intensity-and-labour-quality-sector-level-new-zealand-and-uk-wp-07-01-html#section-4 (accessed 6 February 2019).

58 'Annual national accounts show housing investment has record high share', Stats NZ, www.stats.govt.nz/news/annual-national-accounts-show-housing-investment-has-record-high-share (accessed 21 January 2019).

59 Rachel Griffith, Stephen Redding and John Van Reenen, 'Mapping the two faces of R&D: Productivity growth in a panel of OECD industries', *Review of Economics and Statistics*, 86, 4 (2004), pp.883–95; Bronwyn H. Hall, Jacques Mairesse and Pierre Mohnen, 'Measuring the Returns to R&D', in Bronwyn Hall and Nathan Rosenberg (eds), *Handbook of the Economics of Innovation*, vol.2, North-Holland, Amsterdam, 2010, pp.1033–82.

60 'Research and development in New Zealand: 2016', Stats NZ, 2017, http://archive.stats.govt.nz/browse_for_stats/businesses/research_and_development/research-development-nz-2016.aspx (accessed 16 December 2018); OECD, *Main Science and Technology Indicators 2018/1*, OECD Publishing, Paris, 2018, p.30.

61 R. Green, R. Agarwal, P. Brown, H. Tan and K. Randhawa, 'Management Matters in New Zealand: How Does Manufacturing Measure Up?', Ministry of Economic Development Occasional Paper 11/03, Wellington, 2011.

62 'How Americans Spend Their Money', *New York Times*, 10 February 2008, https://archive.nytimes.com/www.nytimes.com/imagepages/2008/02/10/opinion/10op.graphic.ready.html (accessed 15 December 2018). On close inspection this graph does not show particularly clearly that adoption has been speeding up, despite what the title claims.

63 This is measured as technology adoption as a percentage of the population; see James Manyika, Susan Lund, Michael Chui, Jacques Bughin, Jonathan Woetzel, Parul Batra, Ryan Ko and Saurabh Sanghvi, *Jobs Lost, Jobs Gained: Workforce Transitions in a Time of Automation*, McKinsey Global Institute, 2017, p.50, www.mckinsey.com/~/media/mckinsey/featured%20insights/future%20of%20organizations/what%20the%20future%20of%20work%20will%20mean%20for%20jobs%20skills%20and%20wages/mgi-jobs-lost-jobs-gained-report-december-6-2017.ashx (accessed 30 January 2019).

64 Author interview with Comin.

65 Diego Comin and Bart Hobijn, 'An Exploration of Technology Diffusion', *American Economic Review*, 100, 5 (2010), pp.2031–59.

66 Frey and Osborne, 'The future of employment'.

67 Ashley Carman, 'Juicero, maker of the doomed $400 internet-connected juicer, is shutting down', *The Verge*, 1 September 2007, www.theverge.com/2017/9/1/16243356/juicero-shut-down-lay-off-refund (accessed 15 December 2018).

68 Author interview with Comin; see also Diego Comin and Martí Mestieri, 'If Technology Has Arrived Everywhere, Why Has Income Diverged?', *American Economic Journal: Macroeconomics*, 10, 3 (2018), pp.137–78.

69 Manyika et al., *Jobs Lost, Jobs Gained*, p.50.

70 Paul David, 'The Dynamo and the Computer: An Historical Perspective on the Modern Productivity Paradox', *American Economic Review*, 80, 2 (1990), pp.355–61; Tim Harford, 'Why didn't electricity immediately change manufacturing?', BBC, 21 August 2017, https://www.bbc.com/news/business-40673694 (accessed 31 March 2019).

71 Tim Harford, *Fifty Inventions That Shaped the Modern Economy*,

Riverhead Books, New York, 2017.

72 Brian De La Rue, 'Is robotic milking really the smart option?', Stuff, 2 May 2017, www.stuff.co.nz/business/farming/dairy/92089530/is-robotic-milking-really-the-smart-option (accessed 16 December 2018).

73 Keith Woodford, Marc Brakenrig and Marvin Pangborn, 'New Zealand case studies of automatic-milking-systems adoption', *Proceedings of the New Zealand Society of Animal Production*, 75 (2015), pp.127–31.

74 De La Rue, 'Is robotic milking really the smart option?'

75 Ajay Agrawal, Joshua Gans and Avi Goldfarb, *Prediction Machines: The Simple Economics of Artificial Intelligence*, Harvard Business Review Press, Boston, 2018.

76 'Of prediction and policy', *The Economist*, 20 August 2018, www.economist.com/finance-and-economics/2016/08/20/of-prediction-and-policy (accessed 15 December 2018).

77 De La Rue, 'Is robotic milking really the smart option?'

78 *The New Zealand Vehicle Fleet: Fact and Fiction*, Ministry of Transport, 2017, p.15, www.transport.govt.nz/assets/Uploads/Research/Documents/6595737381/NZ-Vehicle-fleet-fact-and-fiction-2017.pdf (accessed 16 December 2018).

79 Nicky Woolf, 'Elon Musk Twitter rant a "case study" in how not to handle a crisis, experts say', *Guardian*, 7 July 2016, www.theguardian.com/technology/2016/jul/07/tesla-elon-musk-autopilot-death-crisis-management (accessed 15 December 2018).

80 Maeve Henchion, Maria Hayes, Anne Maria Mullen, Mark Fenelon and Brijesh Tiwari, 'Future Protein Supply and Demand: Strategies and Factors Influencing a Sustainable Equilibrium', *Foods*, 6, 7 (2017), https://doi.org/10.3390/foods6070053 (accessed 8 February 2019).

81 Fredrik Erixon and Björn Weigel, *The Innovation Illusion: How So Little Is Created by So Many Working So Hard*, Yale University Press, New Haven, 2016, p.185.

Chapter Three

1 Jeevan Vasagar, 'How robots are teaching Singapore's kids', *Financial Times*, 13 July 2017, www.ft.com/content/f3cbfada-668e-11e7-8526-7b38dcaef614 (accessed 15 December 2018).

2 Larry Hardesty, 'Computer system transcribes words users "speak silently"', *MIT News*, 4 April 2018, http://news.mit.edu/2018/computer-system-transcribes-words-users-speak-silently-0404 (accessed 15 December 2018).

3 Tom Pullar-Strecker, 'Robots develop at speed as they run and incorporate muscle', Stuff, 5 June 2018, www.stuff.co.nz/business/industries/104273427/robots-develop-at-speed-as-they-run-and-incorporate-muscle (accessed 15 December 2018).

4 Laurence Kubiak, Mark Dean and Aaron Drew, 'Disruptive Technologies Risks and Opportunities – Can New Zealand make the most of them?', NZIER, 21 October 2015, https://nzier.org.nz/static/media/filer_public/6d/6e/6d6ecf8b-032c-4551-b0a7-8cd0f39e2004/disruptive_technologies_for_caanz.pdf (accessed 15 December 2018).

5 Gareth Kiernan, *From Education to Employment: Megatrends affecting NZ's working environment*, Infometrics, 2017, p.9, http://static.infometrics.co.nz/Content/Megatrends_web%20version.pdf (accessed 6 February 2019).

6 See John Hawksworth, Richard Berriman and Saloni Goel, *Will Robots Really Steal Our Jobs?*, PricewaterhouseCoopers, 2018, www.pwc.co.nz/pdfs/2018pdfs/impact-of-automation-on-jobs-Feb-2018.pdf (accessed 14 January 2019).

7 To their credit, PwC are more up-

front about the limitations of the research than others, even including it in the executive summary.

8 See, among many others, Martin Ford, *Rise of the Robots: Technology and the Threat of a Jobless Future*, Basic Books, New York, 2015.

9 Chris Weller, 'Elon Musk doubles down on universal basic income: "It's going to be necessary"', *Business Insider*, 13 February 2017, www.businessinsider.com/elon-musk-universal-basic-income-2017-2 (accessed 15 December 2018).

10 'Mark Zuckerberg's Commencement address at Harvard', *Harvard Gazette*, 27 May 2017, https://news.harvard.edu/gazette/story/2017/05/mark-zuckerbergs-speech-as-written-for-harvards-class-of-2017 (accessed 15 December 2018).

11 Chris Weller, '8 high-profile entrepreneurs who have endorsed universal basic income', *Business Insider*, 9 November 2016, www.businessinsider.com/entrepreneurs-endorsing-universal-basic-income-2016-11 (accessed 15 December 2018).

12 'Unemployment rate falls to 3.9 per cent', Stats NZ, 6 November 2018, www.stats.govt.nz/news/unemployment-rate-falls-to-3-9-per cent (accessed 15 December 2018).

13 Colin Williscroft, 'Need a job? Move to Ashburton, it has plenty', Stuff, 1 December 2018, www.stuff.co.nz/the-press/news/108901805/need-a-job-move-to-ashburton-they-have-plenty (accessed 15 December 2018).

14 Mamta Badkar, Sam Fleming and Joe Rennison, 'US unemployment rate hits lowest level since 1969', *Financial Times*, 5 October 2018, www.ft.com/content/e77ae0bc-c88e-11e8-ba8f-ee390057b8c9 (accessed 15 December 2018).

15 Ben Casselman, 'As Labor Pool Shrinks, Prison Time Is Less of a Hiring Hurdle', *New York Times*, 13 January 2018, www.nytimes.com/2018/01/13/business/economy/labor-market-inmates.html (accessed 15 December 2018).

16 Theophilos Argitis, 'Canada's Unemployment Rate Drops to Lowest in Four Decades', Bloomberg, 5 January 2018, www.bloomberg.com/news/articles/2018-01-05/canada-s-unemployment-rate-declines-to-lowest-in-four-decades (accessed 15 December 2018).

17 Gavin Jackson, 'UK unemployment rate falls to lowest since 1975', *Financial Times*, 19 March 2019, https://www.ft.com/content/02ffd626-4a2b-11e9-bbc9-6917dce3dc62 (accessed 31 March 2019).

18 Unless stated otherwise, labour-force participation rates are from 'Labour force participation rate (indicator)', OECD, 2018, https://doi.org/10.1787/8a801325-en (accessed 15 December 2018).

19 Lawrence Summers, 'Men not at work: Lawrence Summers on America's hidden unemployment', *Financial Times*, 23 September 2016, www.ft.com/content/eb0c4aa4-80b5-11e6-8e50-8ec15fb462f4 (accessed 15 December 2018).

20 Claire Cain Miller, 'The Long-Term Jobs Killer Is Not China. It's Automation', *New York Times*, 21 December 2016, www.nytimes.com/2016/12/21/upshot/the-long-term-jobs-killer-is-not-china-its-automation.html (accessed 15 December 2018; N. Gregory Mankiw, 'Why Aren't More Men Working?', *New York Times*, 15 January 2018, www.nytimes.com/2018/06/15/business/men-unemployment-jobs.html (accessed 15 December 2018).

21 Audrey Breitwieser, Ryan Nunn and Jay Shambaugh, 'The recent rebound in prime-age labor force participation', Brookings, 2 August 2018, www.brookings.edu/blog/up-front/2018/08/02/the-recent-rebound-in-prime-age-labor-force-participation (accessed 15 December 2018).

22 Labour-force participation for ages 15 to 64 has risen between 2000 and 2017 from 67.2 to 76.2 per cent for women and from 83.1 to 85.7 per cent for men. See 'LFS by sex and age – indicators', OECD.Stat, https://stats.oecd.org/Index.aspx?DataSetCode=LFS_SEXAGE_I_R (accessed 15 December 2018).

23 Brian Fallow, 'Brian Fallow: Looking for the job limit', *NZ Herald*, 11 May 2018, www.nzherald.co.nz/business/news/article.cfm?c_id=3&objectid=12049188 (accessed 15 December 2018).

24 'Labour market statistics: September 2018 quarter', Stats NZ, 6 November 2018, www.stats.govt.nz/information-releases/labour-market-statistics-september-2018-quarter (accessed 15 December 2018).

25 See, for example, Australia, Germany, Japan, South Korea, Sweden and the United Kingdom. In France and Canada the employment rate is also very close to an all-time high. As discussed the USA is the important exception. For the OECD as a whole, including the USA, the employment rate has only been tracked since 2005, but is also at its highest level ever. 'Employment rate (indicator)', OECD, 2018, https://doi.org/10.1787/1de68a9b-en (accessed 15 December 2018).

26 'Part-time employment rate (indicator)', OECD, 2018, https://doi.org/10.1787/f2ad596c-en (accessed 15 December 2018).

27 Over the same period we have also become dramatically richer, so we continue to work rather than take time off and earn less. 'Hours worked (indicator)', OECD, 2018, https://doi.org/10.1787/47be1c78-en (accessed 15 December 2018).

28 For statistics here, see 'Robot density rises globally', International Federation of Robotics, https://ifr.org/ifr-press-releases/news/robot-density-rises-globally (accessed 15 December 2018).

29 Moreover, as Larry Summers has pointed out, it is not consistent to argue that technological change has been a big cause of declining labour-force participation but then to argue that productivity growth hasn't accelerated yet because of lags in the uptake of technology. Either it has arrived for both, and the weak productivity performance we see is what we are going to get, or it has not arrived for either and something else is behind declining labour-force participation. See Larry Summers speaking at approximately minute 30 in 'Technology and the Workforce, Panel 1', C-SPAN, www.c-span.org/video/?324436-2/technology-workforce-panel-1&start=923 (accessed 16 February 2019).

30 Greg Kaplan and Sam Schulhofer-Wohl, 'Understanding the long-run decline in interstate migration', *International Economic Review*, 58, 1 (2017), pp.57–94.

31 They use two measures of occupational churn: change in each occupation relative to overall occupational change and the sum of occupational changes when there was an absolute loss in jobs. See Robert D. Atkinson and John Wu, *False Alarmism: Technological Disruption and the U.S. Labor Market, 1850–2015*, Information and Technology and Innovation Foundation, 2017, www2.itif.org/2017-false-alarmism-technological-disruption.pdf (accessed 16 February 2019). Others find the same broad trends. See Steven J. Davis and John Haltiwanger, 'Labor Market Fluidity and Economic Performance', NBER Working Paper 20479, 2014, www.nber.org/papers/w20479.pdf (accessed 5 December 2018).

32 For example, the head of the World Economic Forum, Klaus Schwab, writes that 'the speed of current breakthroughs has no historical precedent. When compared with previous industrial revolutions, the Fourth is evolving at an exponential

rather than a linear pace.' See Klaus Schwab, 'The Fourth Industrial Revolution: What it means, how to respond', World Economic Forum, www.weforum.org/agenda/2016/01/the-fourth-industrial-revolution-what-it-means-and-how-to-respond (accessed 25 December 2018).

33 Steven F. Hipple and Emy Sok, 'Tenure of American Workers', Bureau of Labor Statistics, United States Department of Labor, September 2013, www.bls.gov/spotlight/2013/tenure (accessed 15 December 2018).

34 Author's calculations from New Zealand Household Labour Force Surveys 2008 to 2018. Note that this result holds in New Zealand even when looking at individual age categories such as 35–44, 45–54 and 55–64, suggesting demographic shifts are not an explanation. Infometrics, a New Zealand economic consultancy, suggests that occupational churn may be five times higher in the next two decades than in the past seventeen years. (They do not address whether it has been increasing or decreasing by their measure through these seventeen years.) They do not include their methodology for forecasting occupational churn, but if their projections are five times what happened in the last seventeen years they may be overestimating what will happen in the next twenty years. See Gareth Kiernan, 'Putting Automation in Perspective', Infometrics, 21 August 2018, www.infometrics.co.nz/putting-automation-perspective (accessed 15 December 2018).

35 'Employment by job tenure intervals – frequency', OECD.Stat, https://stats.oecd.org/Index.aspx?DataSetCode=TENURE_FREQ (accessed 15 December 2018).

36 Paul Krugman, *The Age of Diminished Expectations: U.S. Economic Policy in the 1990s*, MIT Press, Cambridge, Massachusetts, 1997, p.11.

37 For automation to increase multi-factor productivity, not just labour productivity, the cost of producing using machines over the machines' lifetime must be lower than the cost of producing using workers for that time period. Of course, standard economics suggests businesses should (and will) only invest in automation technologies when they believe this is the case.

38 Gustavo Adler, Romain Duval, Davide Furceri, Sinem Kiliç Çelik, Ksenia Koloskova and Marcos Poplawski-Ribeiro, 'Gone with the Headwinds: Global Productivity', IMF Staff Discussion Note 17/04, p.12, www.imf.org/en/Publications/Staff-Discussion-Notes/Issues/2017/04/03/Gone-with-the-Headwinds-Global-Productivity-44758 (accessed 16 February 2019).

39 See Jamie Culling and Hayden Skilling, 'How does New Zealand stack up? A comparison of labour supply across the OECD', *Reserve Bank Bulletin*, 81, 2 (April 2018); Paul Conway, 'Achieving New Zealand's productivity potential', New Zealand Productivity Commission Research Paper 2-16/1, Wellington, 2016, p.18.

40 Catherine Clifford, 'Google CEO: A.I. is more important than fire or electricity', CNBC, 1 February 2018, www.cnbc.com/2018/02/01/google-ceo-sundar-pichai-ai-is-more-important-than-fire-electricity.html (accessed 15 December 2018).

41 Tim Harford, 'We are still waiting for the robot revolution', *Financial Times*, 30 June 2017, www.ft.com/content/4423a404-5c0f-11e7-9bc8-8055f264aa8b (accessed 18 February 2019).

42 See, for example, Jan Hatzius and Kris Dawsey, 'Doing the Sums on Productivity Paradox v2.0', *Goldman Sachs US Economics Analyst*, 15, 30 (2015), www.goldmansachs.com/briefings/multimedia/productivity-paradox-v2.0-revisited.pdf (accessed 16 February 2015); Joel Moykr, 'Secular Stagnation? Not in Your Life', in Coen Teulings and Richard

Baldwin (eds), *Secular Stagnation: Facts, Causes and Cures*, CEPR Press, London, 2014; Martin Feldstein, 'The U.S. Underestimates Growth', *Wall Street Journal*, 18 May 2015, www.wsj.com/articles/the-u-s-underestimates-growth-1431989720 (accessed 16 February 2019); Erik Brynjolfsson and Andrew McAfee, *The Second Machine Age: Work, Progress, and Prosperity in a Time of Brilliant Technologies*, W.W. Norton & Company, New York, 2014. Brynjolfsson has later retreated from the implication that mismeasurement is a strong explanation, in Erik Brynjolfsson, Daniel Rock and Chad Syverson, 'Artificial Intelligence and the Modern Productivity Paradox: A Clash of Expectations and Statistics', NBER Working Paper 24001, 2017, https://pdfs.semanticscholar.org/30d3/715f9ed3f240e5d208db4f6c8ee8792edf91.pdf (accessed 16 December 2018).

43 The argument is that although these are cheap we spend a lot of our valuable time on them, so perhaps they should be valued more highly than they are based on their cost of production. See, for example, the discussion in Gavyn Davies, 'The Internet and the Productivity Slump', *Financial Times*, 3 April 2016, www.ft.com/content/b34e318e-6ffb-332a-96ac-5ce3de2b4e59 (accessed 16 February 2019).

44 Output and productivity growth in 'mismeasured' sectors would have to be many times their measured growth in the data; see Chad Syverson, 'Challenges to Mismeasurement Explanations for the US Productivity Slowdown', *Journal of Economic Perspectives*, 31, 2 (2017), pp.165–86. Syverson also demonstrates there is no relation between the size of a country's information and communications technology sector and the productivity slowdown.

45 Author interview with Daron Acemoglu, 24 April 2018.

46 Robert J. Gordon, *The Rise and Fall of American Growth: The U.S. Standard of Living Since the Civil War*, Princeton University Press, Princeton, 2016. Recent analysis of American patent history suggests something similar. Hyejin Young and co-authors note that in the American patent system, when an innovation is truly novel they need to open a new code to categorise it. For innovations that are recombinations of other ideas or small advances, the patents are placed under an existing code. The opening of new codes accelerated very sharply through the 1800s but began to flatten off sharply in the 1900s and remains quite flat. This suggests that market-creating innovation may also have slowed down as these are probably more likely to be new code categories altogether. See Hyejin Young, Deborah Strumsky, Luis M.A. Bettencourt and José Lobo, 'Invention as a combinatorial process: Evidence from US patents', *Journal of the Royal Society Interface*, 12, 106 (2015), https://royalsocietypublishing.org/doi/full/10.1098/rsif.2015.0272 (accessed 5 December 2018).

47 In this analysis I take inspiration from Larry Summers's discussion of his own grandmother in his review of Gordon's book: Larry Summers, 'Will our children really not know economic growth?', http://larrysummers.com/2016/02/02/will-our-children-really-not-know-economic-growth (accessed 15 December 2018).

48 Michael Mandel and Bret Swanson, *The Coming Productivity Boom: Transforming the Physical Economy with Information*, The Technology CEO Council, 2017, http://www.techceocouncil.org/news/2017/03/28/press_releases/u.s._on_the_cusp_of_a_productivity_boom_according_to_latest_analysis_commissioned_by_technology_ceo_council (accessed 13 March 2019).

49 Ibid. The authors define digital industries as industries where the main output can easily be

provided in digital form and can be delivered anywhere in the world via the internet. Examples include entertainment, publishing, telecoms, search and finance. Physical industries are those whose output is provided mainly in physical form, such as construction, mining, health care, retailing, food services, education and transportation.

50 Moreover, Moore's law may be ending. Moore's law says, simply put, that the processing power of computers will double every two years. That has indeed been the case and has powered remarkable advances in what computers can do, but there is now strong evidence that it is no longer true. This could reduce the speed of advance in frontier computing and various services that rely on it unless other ways of advancing computing power rapidly are found. See Tom Simonite, 'Moore's Law Is Dead. Now What?', *MIT Technology Review*, 13 May 2016, www.technologyreview.com/s/601441/moores-law-is-dead-now-what (accessed 5 December 2018).

51 'Alexa, send up breakfast: Amazon launches Echo for hotels', Stuff, 20 June 2018, www.stuff.co.nz/business/world/104853528/alexa-send-up-breakfast-amazon-launches-echo-for-hotels (accessed 15 December 2018).

52 Steve Lohr, '"The Beginning of a Wave": A.I. Tiptoes into the Workplace', *New York Times*, 5 August 2018, www.nytimes.com/2018/08/05/technology/workplace-ai.html (accessed 15 December 2018).

53 Rachel Thomas, 'Robots could threaten up to half New Zealand's jobs in next 20 years', Stuff, 20 June 2016, www.stuff.co.nz/business/81237965/Robots-could-threaten-up-to-half-New-Zealands-jobs-in-next-20-years (accessed 15 December 2018).

54 For an excellent discussion of the issue, see Richard Baldwin, *The Globotics Upheaval: Globalization, Robotics, and the Future of Work*, Oxford University Press, New York, 2019.

55 Mae Ryan, Cade Metz and Rumsey Raylor, 'How Robot Hands Are Evolving to Do What Ours Can', *New York Times*, 30 July 2018, www.nytimes.com/interactive/2018/07/30/technology/robot-hands.html?rref=collection%2Fbyline%2Fcade-metz (accessed 15 December 2018).

56 Brynjolfsson and McAfee, *The Second Machine Age*.

57 Brynjolfsson et al., 'Artificial Intelligence and the Modern Productivity Paradox'.

58 See 'Are drivers at risk of losing their jobs?', Infometrics, www.infometrics.co.nz/drivers-risk-losing-jobs (accessed 18 February 2019).

59 'How to become a contact centre worker', Ministry of Business, Innovation and Employment, http://occupationoutlook.mbie.govt.nz/service-industries/contact-centre-workers (accessed 15 December 2018).

60 For the positive view, see Rob Stock, 'Talking your way to a career – how a call centre job can get you on your way', Stuff, 1 July 2017, www.stuff.co.nz/business/money/94010038/talking-your-way-to-a-career (accessed 15 December 2018). For the less positive see 'How to become a contact centre worker', Ministry of Business, Innovation and Employment. These projected declines, however, are likely to be in part due to outsourcing rather than purely AI machines taking over.

61 'Solving the Talent Shortage Build, Buy, Borrow and Bridge', Manpower Group New Zealand, 2018, https://downloads.manpowergroup.co.nz/hubfs/MPG_TalentShortageSurvey_Files/assets/MG_TalentShortage_2018_NZ_2.pdf?hsCtaTracking=a4dba5ee-5029-433e-a9ac-8cd7d990b89e%7C53a7f32b-883f-457d-9a37-f9db139936cf (accessed 15 December 2018).

62 Brynjolfsson et al., 'Artificial Intelligence and the Modern Productivity Paradox'.

63 James Surowiecki, 'Robopocalypse Not', *Wired*, August 2017, www.wired.com/2017/08/robots-will-not-take-your-job (accessed 15 December 2018).

64 This refers to data from six major European economies and the USA. See James Remes, James Manyika, Jaana Remes, Jacques Bughin, Jonathan Woetzel, Jan Mischke and Mekala Krishnan, *Solving the Productivity Puzzle: The Role of Demand and the Promise of Digitalization*, McKinsey Global Institute, 2018, www.mckinsey.com/~/media/McKinsey/Featured%20Insights/Meeting%20societys%20expectations/Solving%20the%20productivity%20puzzle/MGI-Solving-the-Productivity-Puzzle-Report-February-22-2018.ashx (accessed 16 February 2019). New Zealand has also seen a decline in productivity growth, as discussed in Chapter Two, with productivity growth for any productivity cycle not exceeding 3 per cent since 1978. See Conway, 'Achieving New Zealand's productivity potential', p.18.

65 Remes et al., *Solving the Productivity Puzzle*.

66 Adler et al., 'Gone with the headwinds: Global productivity'. New Zealand's capital-to-labour ratio has essentially been flat. We also skew a lot of investment into housing rather than into more productive investments. However, the share of investment going to information and communications technology is strong relative to the OECD. See Paul Conway, 'Can the Kiwi Fly? Achieving Productivity Lift-off in New Zealand', *International Productivity Monitor*, 34 (Spring 2018), pp.40–63.

67 The same is true within sectors for individual businesses; more productive businesses should be able to attract resources and grow at the expense of less productive firms. Economists call this allocative efficiency and it can have large effects. See Lisa Meehan, 'Productivity in New Zealand: The role of resource allocation among firms', New Zealand *Economic Papers*, 2019, https://www.tandfonline.com/doi/full/10.1080/00779954.2019.1573846 (accessed 10 March 2019).

68 If this were the case, it would suggest that lower-productivity-growth sectors were using more resources than higher-productivity-growth sectors. However, as lower-productivity sectors are more labour-intensive this would not necessarily suggest that unemployment would rise but would instead point to a world of both low aggregate productivity growth and low unemployment.

69 Remes et al., *Solving the Productivity Puzzle*, p.36.

70 OECD, *The Future of Productivity*, OECD Publishing, Paris, 2015.

71 Ljubica Nedelkoska and Glenda Quintini, 'Automation, skills use and training', OECD Social, Employment and Migration Working Papers 202, OECD Publishing, Paris, 2018, https://doi.org/10.1787/2e2f4eea-en (accessed 16 December 2018).

72 M. Arntz, T. Gregory and U. Zierahn, 'The Risk of Automation for Jobs in OECD Countries: A Comparative Analysis', OECD Social, Employment and Migration Working Papers 189, OECD Publishing, Paris, 2016.

73 Nedelkoska and Quintini, 'Automation, skills use and training'.

74 AI Forum New Zealand, *Artificial Intelligence: Shaping a Future New Zealand*, 2018, p.53, aiforum.org.nz/wp-content/uploads/2018/07/AI-Report-2018_web-version.pdf (accessed 16 February 2019).

75 James Manyika, Susan Lund, Michael Chui, Jacques Bughin, Jonathan Woetzel, Parul Batra, Ryan Ko and Saurabh Sanghvi, *Jobs Lost, Jobs Gained: Workforce Transitions in a Time of Automation*, McKinsey Global Institute, 2017, p.36, www.mckinsey.com/~/media/mckinsey/featured%20

insights/future%20of%20organizations/what%20the%20future%20of%20work%20will%20mean%20for%20jobs%20skills%20and%20wages/mgi-jobs-lost-jobs-gained-report-december-6-2017.ashx (accessed 30 January 2019).

Chapter Four

1 For all historical examples cited here, see Louis Anslow, 'Robots have been about to take all the jobs for more than 200 years', Timeline, 16 May 2016, https://timeline.com/robots-have-been-about-to-take-all-the-jobs-for-more-than-200-years-5c9c08a2f41d (accessed 17 December 2018).

2 David H. Autor, 'Why Are There Still So Many Jobs? The History and Future of Workplace Automation', *Journal of Economic Perspectives*, 29, 3 (2015), p.4.

3 Anslow, 'Robots have been about to take all the jobs for more than 200 years'.

4 This model is based on Acemoglu and Restrepo's model set out in the following paper: Daron Acemoglu and Pascual Restrepo, 'Artificial Intelligence, Automation and Work', in Ajay Agarwal, Avi Goldfarb and Joshua Gans (eds), *Economics of Artificial Intelligence: An Agenda*, University of Chicago Press, Chicago, forthcoming, www.nber.org/chapters/c14027.pdf (accessed 29 January 2019). They include a capital accumulation effect, which is also good for workers; for the sake of simplicity and clarity I have left that out of the discussion here but the article provides full details. These ideas are also developed in Daron Acemoglu and Pascual Restrepo, 'The Race between Man and Machine: Implications of Technology for Growth, Factor Shares, and Employment', *American Economic Review*, 108, 6 (2018), pp.1488–542, and Daron Acemoglu and Pascual Restrepo, 'Automation and New Tasks: How Technology Displaces and Reinstates Labor', NBER Working Paper 25684, March 2019, https://www.nber.org/papers/w25684 (accessed 4 April 2019).

5 In referring to wages, I broadly mean wages, salaries and other labour income.

6 This is what economists refer to as a rise in real wages.

7 This famous moment in American television is available at 'Oprah you get a car', YouTube, www.youtube.com/watch?v=hcJAWKdawuM (accessed 17 December 2018).

8 The extent to which we see the home-sector effect vs the other-sector effect depends on the price elasticity of demand, i.e., the change in how much of a good people demand when the price of that good changes. If demand is highly elastic the home-sector effect is likely to be stronger as people buy a lot more of that good when the price drops; if it is less elastic then consumers will not buy much more (or even in some rare cases no more or less), and will use the savings from this good on another sector instead, meaning the other-sector effect is stronger. Regardless of which effect dominates, there is an effect in a positive direction for workers overall – either because more of the home sector's goods are consumed, helping the remaining workers in that industry, or goods from another sector are consumed, helping the workers in other industries.

9 This book will largely consider the other-sector effect via the channel of goods being cheaper and thus households having more to spend on products from other sectors, which increases the need for workers in those other sectors. However, there is also likely to be an employment effect for industries that are linked closely to the automating sector. Businesses that produce inputs or use the outputs of the automating sector may see greater demand for their goods thanks to cheaper production in the automating sector spurring

demand. For example, automation in postal and delivery services would make them cheaper, and would in turn make all the products sold via delivery cheaper, potentially stimulating demand for a whole range of products that are commonly delivered. This would contribute to employment growth in the businesses that make these products. See David Autor and Anna Salomons, 'Is Automation Labor-Displacing? Productivity Growth, Employment, and the Labor Share', NBER Working Paper 24871, 2018, www.nber.org/papers/w24871 (accessed 17 December 2018).

10 For further discussion of why this can be the case see Acemoglu and Restrepo, 'The Race between Man and Machine'.

11 Workers of course need to have the right skills to participate in these new tasks. This challenge is discussed further in Chapters Five and Seven.

12 Eleanor Catton, *The Luminaries,* Granta, London, 2013, p.125.

13 'The Mining Process', Oceana Gold, www.waihigold.co.nz/mining/the-mining-process (accessed 15 December 2018).

14 James Bessen, *Learning by Doing: The Real Connection between Innovation, Wages, and Wealth,* Yale University Press, New Haven, 2015, pp.96–97.

15 Ibid.

16 This section draws on the author's interview with James Bessen on 23 April 2018. See also James Bessen, 'How Computer Automation Affects Occupations: Technology, Jobs, and Skills', Boston University School of Law, Law and Economics Research Paper 15-49, 2016, https://papers.ssrn.com/sol3/papers.cfm?abstract_id=2690435 (accessed 17 December 2018).

17 Ibid.

18 James Manyika, Michael Chui, Mehdi Miremadi, Jacques Bughin, Katy George, Paul Willmott and Martin Dewhurst, *A Future that Works: Automation, Employment, and Productivity,* McKinsey Global Institute, 2017, p.1, www.mckinsey.com/~/media/McKinsey/Featured%20Insights/Digital%20Disruption/Harnessing%20automation%20for%20a%20future%20that%20works/MGI-A-future-that-works_Full-report.ashx (accessed 3 February 2019).

19 Normally, cheaper prices mean people on average buy more of a good or service. In some rare cases, a decline in the price of a good may mean consumers demand *less* of it because they are wealthier and can therefore choose to buy different goods altogether. In other cases, particularly if the good is already cheap, consumers may simply not demand much more of it despite the price drop, because they prefer to spend the money they've saved on things from other sectors. Yet in both these cases the other-sector productivity effect (which would help increase demand for workers in other sectors) would dominate.

20 The key condition for this to occur is that the price elasticity of demand needs to be sufficiently large – in other words people must buy a lot more of the product when it becomes cheaper.

21 James Bessen, 'Toil and Technology', *Finance and Development,* 52, 1 (2015), pp.16–19.

22 Timothy Lee, 'Why history suggests that today's wage stagnation is temporary', *Vox,* 21 May 2015, www.vox.com/2015/5/21/8630771/software-slow-wage-growth (accessed 17 December 2018).

23 For example Autor and Salomons find that, over time, employment in industries with high productivity growth declines, suggesting that the home-sector productivity effect eventually runs out of steam. However, they also find that the declines in employment in sectors with high automation are more than offset by employment growth

in other sectors. This comes from employment growth in the economy as a whole thanks to greater aggregate demand, but also from employment growth sectors that are downstream customers of the sector experiencing high productivity growth, a more direct effect. This is a combination of the other-sector productivity effect and the new-task effect. They also note that the declines in labour share in high-productivity sectors are not offset elsewhere. The decline of labour share and its possible causes are discussed in more detail later in this chapter. See Autor and Salomons, 'Is Automation Labor-Displacing?'

24 In economics terms, at this point the demand is inelastic as further reductions in price do not generate any extra purchases.

25 For the study on the American Midwest, see Daron Acemoglu and Pascual Restrepo, 'Robots and Jobs: Evidence from US Labor Markets', NBER Working Paper 23285, March 2017, https://www.nber.org/papers/w23285.pdf (accessed 11 March 2019); for the critique, see Lawrence Mishel and Josh Bivens, 'The zombie robot argument lurches on', Economic Policy Institute, 24 May 2017, www.epi.org/files/pdf/126750.pdf (accessed 15 December 2018).

26 Georg Graetz and Guy Michaels, 'Robots at Work', *Review of Economics and Statistics,* 100, 5 (December 2018), www.mitpressjournals.org/doi/abs/10.1162/rest_a_00754 (accessed 15 January 2019). Their work does include some suggestive evidence that lower-skill workers might lose out but notes that overall employment does not decline. These distributional impacts, even if minor, need careful handling, as Chapter Five suggests. In another paper in the *American Economic Review* they note that while 'jobless recoveries' (economic recoveries from recessions that include GDP growth but little employment growth) have been common in the USA, these are far less common in other developed countries, and there is little evidence that sectors using more modern technology had more trouble generating jobs in recovery than other sectors. Technology in short, does not seem to be to blame. Again, the policy environment in other countries compared to the USA may play an important role. Notably, Europe also has substantially more industrial robots per thousand workers than the USA so in fact the opposite result might be expected. See Georg Graetz and Guy Michaels, 'Is Modern Technology Responsible for Jobless Recoveries?', *American Economic Review,* 107, 5 (2017), pp.168–73.

27 Indeed, at least in the short run, various microeconomic studies, including Graetz and Michaels, and Alexopoulos and Cohen (who look at the twentieth century), do not find a decline in employment in sectors affected by technology. See Michelle Alexopoulos and Jon Cohen, 'The Medium Is the Measure: Technical Change and Employment, 1909–1949', *Review of Economics and Statistics,* 98, 4 (2016), pp.792–810. However, as mentioned in a previous footnote, Autor and Salomons in recent work find that over time higher productivity sectors eventually see decline in employment. However, another study by Salomons finds that once you factor in the broader effects on employment in the economy (not just in the sector directly affected by automation), the introduction of automation technologies is a net positive for employment even in the longer run. She shows that, overall, technology that replaced routine work did not reduce employment in Europe, and instead increased it by 11.6 million jobs, half of the total rise observed in the period between 1999 and 2010. This is significantly because consumers spent more money on products that were now cheaper thanks to automation, raising overall labour demand for those

sectors. She also finds that increased wealth through non-wage income is particularly important to supporting this job creation. See Terry Gregory, Anna Salomons and Ulrich Zierahn, 'Racing With or Against the Machine? Evidence from Europe', Utrecht School of Economics, Tjalling C. Koopmans Research Institute, Discussion Paper Series 16-05, 2016, http://ftp.zew.de/pub/zew-docs/dp/dp16053.pdf (accessed 17 December 2018).

28 This comes primarily because products are cheaper – which is a benefit that everyone who spends on products gets regardless of their income source. However, it can be stronger still if workers also have non-wage income, like shares in companies that are using automation technologies, which would also see them increase their income directly.

29 Holding roughly steady the product, this same fact should be true for any sector that has seen significant productivity growth, and is therefore particularly true for manufactured goods. As *The Economist* points out, 'In 1908 the average American had to work for around 4700 hours to buy a Model T Ford. A century later a typical car can be had for only 1365 hours of labour.' See 'An incurable disease', *The Economist*, 29 September 2012, www.economist.com/finance-and-economics/2012/09/29/an-incurable-disease (accessed 17 December 2018). See also Annie Lowrey, 'Changed Life of the Poor: Better Off, but Far Behind', *New York Times*, 30 April 2014, www.nytimes.com/2014/05/01/business/economy/changed-life-of-the-poor-squeak-by-and-buy-a-lot.html (accessed 15 December 2018).

30 While this is true as a share of our income, it is possible for the amount spent on productive sectors to keep growing in absolute terms; it's just that spending on less productive sectors is likely to grow faster still.

31 We could choose to save more too, but eventually this will be invested or spent in the economy.

32 This is mirrored by the price of goods declining relative to services. See Mark J. Perry, 'Manufacturing's Declining Share of GDP is a Global Phenomenon, and It's Something to Celebrate', U.S. Chamber of Commerce Foundation, 22 March 2012, www.uschamberfoundation.org/blog/post/manufacturing-s-declining-share-gdp-global-phenomenon-and-it-s-something-celebrate/34261 (accessed 15 December 2018).

33 See Autor and Salomons, 'Is Automation Labor-Displacing?' See also Gregory, Salomons and Zierahn, 'Racing With or Against the Machine?'. As is discussed later in the chapter, while employment may remain high thanks to these two effects, the share of income going to labour could still decline as a result of technological change if we rely only on the home- and other-industry productivity effects described here. As per Acemoglu and Restrepo's work, it appears that inventing new tasks is required to keep labour's share of income high.

34 While the supply of land that could be farmed kept growing, this had a very limited impact on employment, but as expanding farmland became harder, the fact one farmer could do more each day did displace some workers.

35 'New Zealand Fitness Industry Overview', New Zealand Institute of Health and Fitness, https://nzihf.ac.nz/personal-training-careers/new-zealand-fitness-industry-overview (accessed 15 December 2018). This wasn't the only way jobs were created; Les Mills has also started exporting the workouts and concepts. Their eighteen fitness programmes, virtual programmes and equipment solutions can be found in 15,500 clubs in eighty countries.

36 New tasks don't always map perfectly to new jobs, but many new tasks generate new jobs.

37 Laurel Funk, '30 Jobs That Didn't Exist 30 Years Ago', GoBankingRates, 3 March 2018, www.gobankingrates.com/making-money/jobs/jobs-didnt-exist-30-years-ago/#0 (accessed 15 December 2018); Kyriaki Raouna, '12 Jobs That Didn't Exist 10 Years Ago', Career Addict, 19 August 2016, www.careeraddict.com/12-jobs-that-didn-t-exist-10-years-ago (accessed 15 December 2018); Emily Moore, '15 Awesome Jobs That Didn't Exist 15 Years Ago', Glassdoor, 23 October 2017, www.glassdoor.com/blog/jobs-that-didnt-exist-15-years-ago (accessed 15 December 2018).

38 Acemoglu and Restrepo, 'The Race between Man and Machine'.

39 See Autor and Salomons, 'Is Automation Labor-Displacing?' and Gregory, Salomons and Zierahn, 'Racing With or Against the Machine?'. See also David Autor, David Dorn and Gordon Hanson, 'Untangling Trade and Technology: Evidence from Local Labour Markets', *Economic Journal*, 125, 584 (2015), pp.621–46.

40 Paul Callister and Robert Didham, 'Workforce composition – Workforce by sector', Te Ara – the Encyclopedia of New Zealand, https://teara.govt.nz/en/workforce-composition/page-1 (accessed 15 December 2018).

41 This is not to say there is no hunger problem in New Zealand; however, this is not a result of inadequate food supplies – we export large amounts of food. It is instead a problem of poverty and access to the food we do have.

42 Carl Benedikt Frey and Michael A. Osborne, 'The future of employment: How susceptible are jobs to computerisation?', *Technological Forecasting and Social Change*, 114 (2017), pp.254–80.

43 Population growth will account for the remaining 25 per cent of demand increase. See James Manyika, Susan Lund, Michael Chui, Jacques Bughin, Jonathan Woetzel, Parul Batra, Ryan Ko and Saurabh Sanghv, *Jobs Lost, Jobs Gained: Workforce Transitions in a Time of Automation*, McKinsey Global Institute, 2017, p.57, www.mckinsey.com/~/media/mckinsey/featured%20insights/future%20of%20organizations/what%20the%20future%20of%20work%20will%20mean%20for%20jobs%20skills%20and%20wages/mgi-jobs-lost-jobs-gained-report-december-6-2017.ashx (accessed 30 January 2019).

44 'Digital Nation New Zealand – From Tech Sector to Digital Nation', New Zealand Technology Industry Association, August 2016, https://nztech.org.nz/resources/about-the-sector (accessed 15 December 2018).

45 Eleanor Wenman, 'NZ needs more than 50,000 construction workers by 2022 to keep up with demand', Stuff, 2 May 2018, www.stuff.co.nz/business/industries/103495394/nz-needs-more-than-50000-construction-workers-by-2022-to-keep-up-with-demand (accessed 15 December 2018).

46 *World Employment and Social Outlook: Trends 2018*, International Labour Organization, Geneva, 2018, p.44.

47 'Renewable energy resources', Energy Efficiency and Conservation Authority, www.eeca.govt.nz/energy-use-in-new-zealand/renewable-energy-resources (accessed 17 December 2018).

48 87 per cent of New Zealand three-year-olds go to preschool, above the OECD average but still less than in eleven other countries, which suggests increases in early-childhood-care demand are likely with related rises in employment. 'Starting Strong IV: Early Childhood Education and Care Data Country Note', OECD, www.oecd.org/education/school/ECECDCN-NewZealand.pdf (accessed 15 December 2018).

49 'LFS by sex and age – indicators', OECD.Stat, https://stats.oecd.org/Index.aspx?DataSetCode=LFS_SEXAGE_I_R (accessed 15 December 2018).

50 Manyika et al., *Jobs Lost, Jobs Gained*.

51 Developing economies, they suggest, will be less affected by automation and rising unemployment.

52 *Asian Development Outlook 2018: How Technology Affects Jobs*, Asian Development Bank, 2018, p.78, http://dx.doi.org/10.22617/FLS189310-3 (accessed 28 January 2019).

53 See Dan Shewan, 'Robots will destroy our jobs – and we're not ready for it', *Guardian*, 11 January 2017, www.theguardian.com/technology/2017/jan/11/robots-jobs-employees-artificial-intelligence (accessed 15 December 2018); Nicholas Agar, 'Don't trust economists about the future of work', Future Learning, 18 September 2017, https://www.newsroom.co.nz/@future-learning/2017/09/17/48142/dont-trust-economists-about-the-future-of-work (accessed 19 March 2019).

54 Andrew Lo, *Adaptive Markets, Financial Evolution at the Speed of Thought*, Princeton University Press, Princeton, 2017, p.85.

55 For a detailed discussion see Steven Pinker, *Enlightenment Now: The Case for Reason, Science, Humanism, and Progress*, Viking, New York, 2018.

56 Brian Easton, 'Unemployment, 1896–2006', Te Ara – the Encyclopedia of New Zealand, https://teara.govt.nz/en/graph/24362/unemployment-1896-2006 (accessed 15 January 2019).

57 Thomas Manch, 'New Zealand's "working poor" and the push to understand how many are struggling', Stuff, 16 May 2018, www.stuff.co.nz/national/103761074/new-zealands-working-poor-and-the-push-to-understand-how-many-are-struggling (accessed 3 February 2019).

58 The minimum wage has since risen slightly.

59 See Manch, 'New Zealand's "working poor" and the push to understand how many are struggling', and Anna Bracewell-Worrall, 'Official briefings paint bleak picture of NZ's working poor', Newshub, 8 December 2017, www.newshub.co.nz/home/politics/2017/12/official-briefings-paint-bleak-picture-of-nz-s-working-poor.html (accessed 17 December 2018).

60 IMF, 'Understanding the Downward Trend in Labor Income Shares', in *World Economic Outlook, April 2018: Cyclical Upswing, Structural Change*, IMF, Washington, DC, 2018, ch.3.

61 'Does New Zealand have low wages? Yes', *CTU Monthly Economic Bulletin*, 202 (August 2018), p.6, www.union.org.nz/wp-content/uploads/2018/09/CTU-Monthly-Economic-Bulletin-202-August-2018.pdf (accessed 15 December 2018). Although there are some disagreements on the exact size of the fall, Huon Fraser at the New Zealand Productivity Commission in its latest update suggests the decline for core New Zealand industries since 1978 was 8.3 percentage points. Recent declines have been more modest in his view, with a decline of about 1 per cent between 2010 and 2016. See Huon Fraser, 'The Labour Income Share in New Zealand: An Update', New Zealand Productivity Commission Staff Research Note 2018/1, 2018.

62 The 2018 figure comes from the author's email correspondence with Bill Rosenberg on 1 February 2019. These figures have been calculated in the past in a similar fashion; see Jenn Lawless, 'Workers out of pocket $11,500 a year as a result of decades of poor government policy', Council of Trade Unions, 22 March 2018, www.union.org.nz/workers-out-of-pocket-11500-a-year-as-a-result-of-decades-of-poor-government-policy (accessed 17 December 2018).

63 In economics parlance, the elasticity of substitution between capital and labour has to be more than 1.

64 'Technology and the decline in the U.S. labor share of income', Washington Center for Equitable Growth, 15 January 2015, https://equitablegrowth.org/technology-decline-u-s-labor-share-income (accessed 15 December 2018). See

also Loukas Karabarbounis and Brent Neiman, 'The Global Decline of the Labor Share', NBER Working Paper 19136, June 2013, www.nber.org/papers/w19136 (accessed 17 December 2018), and also Robert Z. Lawrence, 'Explaining Recent Declines in Labor's Share in US Income', Peterson Institute for International Economics, 15 October 2015, https://piie.com/commentary/op-eds/explaining-recent-declines-labors-share-us-income (accessed 15 December 2018).

65 Cardiff Garcia, 'Technology vs globalisation as cause of the declining labour share', *Financial Times*, 12 December 2013, https://ftalphaville.ft.com/2013/12/12/1718312/technology-vs-globalisation-as-cause-of-the-declining-labour-share (accessed 15 December 2018).

66 Autor and Salomons, 'Is Automation Labor-Displacing?'

67 Michael Elsby, Bart Hobijn and Aysegul Sahin, 'The Decline of U.S. Labour Share', Brookings Papers on Economic Activity, Fall 2013, www.brookings.edu/wp-content/uploads/2016/07/2013b_elsby_labor_share.pdf (accessed 17 December 2018).

68 Peter Lyons, 'Peter Lyons: Why our most profitable industries pay miserable wages', *NZ Herald*, 5 June 2018, www.nzherald.co.nz/business/news/article.cfm?c_id=3&objectid=12064199 (accessed 17 December 2018).

69 Bill Rosenberg, 'A brief history of labour's share of income in New Zealand 1939–2016', in Gordon Anderson (ed.), *Transforming Workplace Relations*, Victoria University Press, Wellington, 2017, p.99.

70 See Gabriele Ciminelli, Romain Duval and Davide Furceri, 'Employment Protection Deregulation and Labor Shares in Advanced Countries', IMF Working Paper 18/186, 2018, www.imf.org/en/Publications/WP/Issues/2018/08/16/Employment-Protection-Deregulation-and-Labor-Shares-in-Advanced-Economies-46074 (accessed 29 January 2019). See also Lawrence Mishel, 'Causes of Wage Stagnation', Economic Policy Institute, 6 January 2015, www.epi.org/publication/causes-of-wage-stagnation (accessed 17 December 2018).

71 Gauti Eggertson, Jacob Robbins and Ella Getz Wold, 'Kaldor and Piketty's facts: The rise of monopoly power in the United States', Washington Center for Equitable Growth Working Paper Series, 12 February 2018, https://equitablegrowth.org/wp-content/uploads/2018/02/02052018-WP-kaldor-piketty-monopoly-power.pdf (accessed 17 December 2018).

72 See Acemoglu and Restrepo, 'Automation and New Tasks'. They find that new task creation is critical to maintaining a high labour share of income and labour demand, but that in the USA new task creation may have slowed down in recent decades. They blame policy incentives that favour innovations that automate work (efficiency innovations) over innovations that create new tasks (market-creating innovations). Change policy, they suggest, and the picture could improve. They also point out that the creation of new tasks can improve multi-factor productivity by allowing labour to shift to these new tasks in which it has a comparative advantage, meaning labour and capital are both used more efficiently.

73 Anna Stansbury and Lawrence H. Summers, 'Productivity and Pay: Is the link broken?', NBER Working Paper 24165, 2017, www.nber.org/papers/w24165 (accessed 3 February 2019).

74 'Does New Zealand have low wages? Yes', *CTU Monthly Economic Bulletin*.

75 Rana Foroohar, 'Vivienne Ming: "The professional class is about to be blindsided by AI"', *Financial Times*, 27 July 2018, www.ft.com/content/3aac2330-8f38-11e8-b639-

7680cedcc421 (accessed 15 December 2018).

76 Richard Susskind and Daniel Susskind, *The Future of the Professions: How Technology Will Transform the Work of Human Experts*, Oxford University Press, Oxford, 2015.

77 Eric Hobsbawm, 'The Machine Breakers', *Past and Present*, 1 (February 1952), p.59.

78 David Carey, 'Adapting to the changing labour market in New Zealand', OECD Economics Department Working Paper 1420, OECD Publishing, Paris, 2017.

79 Author interview with David Autor, 23 April 2018.

80 David Autor and David Dorn, 'The Growth of Low-Skill Service Jobs and the Polarization of the US Labor Market', *American Economic Review*, 103, 5 (2013), pp.1553–97.

81 Maarten Goos, Alan Manning and Anna Salomons, 'Job Polarization in Europe', *American Economic Review*, 99, 2 (2009), pp.58–63.

82 OECD, *OECD Economic Surveys: New Zealand 2017*, OECD Publishing, Paris, 2017, p.45.

83 Goos, Manning and Salomons, 'Job Polarization in Europe'.

84 OECD, *OECD Economic Surveys: New Zealand 2017*, p.45. High-education occupations typically require tertiary education. They include teaching professionals, other professionals, life-science professionals, science professionals, life-science associate professors, public administration, science associate professors, other associate professors and corporate managers. Middle-education occupations typically call for upper-secondary-school attainment. They include office clerks, customer service, building trades, metal and machinery trades, precision trades, agriculture and fisheries personal services, salespersons and other trades. Low-education occupations have limited educational requirements. These occupations include industrial plant operators, building and related occupations, machine operators, elementary occupations and drivers.

85 Author's calculations using 'Persons Employed by Sex by Occupation', Household Labour Force Survey 2011–2017, Stats NZ, http://archive.stats.govt.nz/infoshare (accessed 6 February 2019).

86 David Rotman, 'Technology and Inequality', *MIT Technology Review*, 21 October 2014, www.technologyreview.com/s/531726/technology-and-inequality (accessed 15 December 2018).

87 Lawrence Mishel and Julia Wolfe, 'Top 1.0 per cent reaches highest wages ever – up 157 per cent since 1979', Economic Policy Institute, 18 October 2018, www.epi.org/blog/top-1-0-per cent-reaches-highest-wages-ever-up-157-per cent-since-1979 (accessed 17 December 2018).

88 Orsetta Causa and Mikkel Hermansen, 'Income redistribution through taxes and transfers across OECD countries', OECD Economics Department Working Paper 1453, OECD Publishing, Paris, 2017, p.20.

89 See Lawrence Mishel and Jessica Schieder, 'CEO compensation surged in 2017', Economic Policy Institute, August 2018, www.epi.org/files/pdf/152123.pdf (accessed 15 December 2018). For a detailed discussion see Deborah Hargreaves, *Are Chief Executives Overpaid?*, Polity Press, Cambridge, UK, 2018.

90 'New Zealand Pay Ratios', Otago Business School, www.otago.ac.nz/business/research/department/otago119826.html (accessed 15 December 2018).

91 Mishel and Bivens, 'The zombie robot argument lurches on'.

92 OECD, *The Productivity-Inclusiveness Nexus*, OECD Publishing, Paris, 2018.

93 These two are profiled in 'The Big Read: Are you ready for the new "gig economy"?', *NZ Herald*, 26 September 2015, www.nzherald.co.nz/business/news/article.cfm?c_

id=3&objectid=11519509 (accessed 15 December 2018).

94 See finding that 25 per cent of 2.8 million gig-economy workers earn below the minimum hourly wage in Katriina Lepänjuuri, Robert Wishart and Peter Cornick, 'The characteristics of those in the gig economy, Final Report', Department of Business Energy and Industrial Strategy, February 2018, p.6, https://assets.publishing.service.gov.uk/government/uploads/system/uploads/attachment_data/file/687553/The_characteristics_of_those_in_the_gig_economy.pdf (accessed 15 December 2018).

95 Janine Berg, Marianne Furrer, Ellie Harmon, Uma Rani and M. Six Silberman, *Digital Labour Platforms and the Future of Work: Towards Decent Work in the Online World*, International Labour Organization, Geneva, 2018, p.xvi.

96 Bruce Munro, 'Struggling for a decent job, decent wage, decent life', *Otago Daily Times*, 24 July 2017, https://www.odt.co.nz/lifestyle/magazine/hours-work (accessed 30 March 2019).

97 Cody Cook, Rebecca Diamond, Jonathan Hall, John A. List and Paul Oyer, 'The Gender Earnings Gap in the Gig Economy: Evidence from over a Million Rideshare Drivers', NBER Working Paper 24732, 2018, www.nber.org/papers/w24732 (accessed 17 December 2018).

98 See James Manyika, Susan Lund, Jacques Bughin, Kelsey Robinson, Jan Mischke and Deepa Mahajan, *Independent Work: Choice, Necessity, and the Gig Economy*, McKinsey Global Institute, 2016, pp.7, 8, www.mckinsey.com/~/media/McKinsey/Featured%20Insights/Employment%20and%20Growth/Independent%20work%20Choice%20necessity%20and%20the%20gig%20economy/Independent-Work-Choice-necessity-and-the-gig-economy-Full-report.ashx (accessed 6 February 2019). Quite a lot depends on the exact definition of independent work. This study looks at workers with: a high degree of autonomy; payment by task, assignment or sales; and a short-term relationship between worker and client. It does not require that the work is through digital platforms. A UK study that defined the gig economy as 'exchange of labour for money between individuals or companies via digital platforms that actively facilitate matching between providers and customers, on a short-term and payment by task basis' found that only 8 per cent saw it as their main source of income. It also found similarly positive attitudes despite often-low pay rates, with people valuing the independence and flexibility it offered. See Lepänjuuri, Wishart and Cornick, 'The characteristics of those in the gig economy'.

99 'Permanently temporary: the Gig Trap', University of Auckland, 10 November 2018, www.auckland.ac.nz/en/news/2018/11/10/permanently-temporary-the-gig-trap.html (accessed 15 December 2018). In the USA, though 60 per cent of Uber drivers seem to move on within six months of their first ride, of course it is not clear if they move to full-time work or to other contracts or gigs. See Cook et al., 'The Gender Earnings Gap in the Gig Economy'.

100 See Louis Thyman, *Temp: How American Work, American Business, and the American Dream Became Temporary*, Viking, New York, 2018.

101 See 'Temporary employment (indicator)', OECD, 2018, https://doi.org/10.1787/75589b8a-en (accessed 11 December 2018). Self-employment has also declined as a share of employment in the European Union and the USA between 2000 and 2017 (data is not available for the OECD as a whole). See 'Self-employment rate (indicator)', OECD, 2019, https://doi.org/10.1787/fb58715e-en (accessed 3 February 2019). Similarly,

in the US the share of independent contractors, which includes many gig-economy workers, declined between 2005 and 2017. See Jay Shambaugh, Ryan Nunn and Lauren Bauer, 'Independent workers and the modern labor market', Brookings, 7 June 2018, www.brookings.edu/blog/up-front/2018/06/07/independent-workers-and-the-modern-labor-market (accessed 3 February 2019).

102 Specifically, there has been a 30 per cent increase since 2010, while the population has only grown about 10 per cent. For a useful discussion, see Jacob McSweeny and Sarah Robson, 'Brighter Future? Beyond 9 to 5', Radio New Zealand, 16 June 2017, https://www.radionz.co.nz/programmes/brighter-future/story/201845837/brighter-future-beyond-9-to-5 (accessed 11 March 2019).

103 'Uber drivers take a day off to highlight low pay', Radio New Zealand, 14 May 2018, www.radionz.co.nz/national/programmes/ninetonoon/audio/2018644787/uber-drivers-take-a-day-off-to-highlight-low-pay (accessed 15 December 2018).

104 'Characteristics of the underemployed in New Zealand', Stats NZ, 24 June 2018, www.stats.govt.nz/reports/characteristics-of-the-underemployed-in-new-zealand (accessed 15 December 2018).

105 See Tim Harford, 'Would basic incomes or basic jobs be better when robots take over?', *Financial Times*, 29 June 2018, www.ft.com/content/c1c8a28e-7a17-11e8-8e67-1e1a0846c475 (accessed 15 December 2018).

Chapter Five

1 In the long run, if enough labour shifts into typically lower-productivity sectors, aggregate productivity may become less of a critical measure of automation, but to get this point we would need to see rapidly rising productivity growth in some major sectors. See Adair Turner, 'Capitalism in the age of robots: Work, income and wealth in the 21st-century', Ineteconomics, 10 April 2018, www.ineteconomics.org/uploads/papers/Paper-Turner-Capitalism-In-The-Age-Of-Robots.pdf (accessed 15 December 2018).

2 Precision agriculture may have other benefits, though, such as reducing the environmental footprint of our agriculture, which may in turn help our competitiveness in global markets likely to be more sensitive to climate change and environmental concerns. As there is limited land on which to produce, reductions in the price of agricultural products thanks to automation may stimulate demand, but are unlikely to cause increases in employment through the home-sector productivity effect, particularly if environmental regulation is strong. This is because land and environmental constraints will limit the extent to which New Zealand can increase production in the sector and therefore the extent of additional employment.

3 This type of agriculture may be more resilient to lab-grown alternatives than mass-produced animal protein.

4 See Kari Leibowitz, Emerson Hardebeck, Parker Goyer and Alia Crum, 'Physician Assurance Reduces Patient Symptoms in US Adults: An Experimental Study', *Journal of General Internal Medicine*, 33, 12 (2018), pp.2051–52. For a broad overview of the research, see Lauren Howe and Kari Leibowitz, 'Can a Nice Doctor Make Treatments More Effective?', *New York Times*, 22 January 2019, www.nytimes.com/2019/01/22/well/live/can-a-nice-doctor-make-treatments-more-effective.html (accessed 29 January 2019). See also Kim Tingley, 'Trying to put a value on the doctor-patient relationship', *New York Times*, 16 May 2018, www.nytimes.com/interactive/2018/05/16/magazine/health-issue-reinvention-of-primary-

care-delivery.html (accessed 15 December 2018).

5 Daniel H. Cohen, 'The new boom in home tuition – if you can pay £40 an hour', *Guardian*, 25 October 2013, www.theguardian.com/education/2013/oct/25/new-boom-home-tuition (accessed 15 December 2018).

6 Timothy Lee, 'The productivity paradox: Why we're getting more innovation but less growth', *Vox*, 24 October 2016, www.vox.com/newmoney/2016/10/24/13327014/productivity-paradox-innovation-growth (accessed 21 January 2019).

7 'Sales and stocks by industry, in current and constant prices (SAFC) (Annual-Mar) Food and beverage services', Retail Trade Survey, Stats NZ, http://archive.stats.govt.nz/infoshare (accessed 18 December 2018).

8 Amanda Cropp, 'More than 2700 new hospitality outlets split the dining dollar', Stuff, 2 September 2018, www.stuff.co.nz/business/106413742/more-than-2700-new-hospitality-outlets-split-the-diner-dollar (accessed 15 December 2018).

9 See David Sax, 'End the Innovation Obsession', *New York Times*, 7 December 2018, www.nytimes.com/2018/12/07/opinion/sunday/end-the-innovation-obsession.html (accessed 15 December 2018). In the New Zealand context, see 'To market, to market – NZ farmers markets', 100% Pure New Zealand, 1 February 2018, https://media.newzealand.com/en/story-ideas/to-market-to-market-nz-farmers-markets (accessed 18 December 2018).

10 Julie Jargon, 'At Starbucks, Baristas Told No More Than Two Drinks', *Wall Street Journal*, 13 October 2010, www.wsj.com/articles/SB10001424052748704164004575548403514060736 (accessed 15 December 2018).

11 James Manyika, Susan Lund, Michael Chui, Jacques Bughin, Jonathan Woetzel, Parul Batra, Ryan Ko and Saurabh Sanghvi, *Jobs Lost, Jobs Gained: Workforce Transitions in a Time of Automation*, McKinsey Global Institute, 2017, p.78, www.mckinsey.com/~/media/mckinsey/featured%20insights/future%20of%20organizations/what%20the%20future%20of%20work%20will%20mean%20for%20jobs%20skills%20and%20wages/mgi-jobs-lost-jobs-gained-report-december-6-2017.ashx (accessed 30 January 2019). For a longer discussion see Tyler Cowen, *Average is Over: Powering America Beyond the Age of the Great Stagnation*, Penguin Group, New York, 2013.

12 See, for example, Claudia Goldin and Lawrence Katz, *The Race between Education and Technology*, Belknap Press, Cambridge, Massachusetts, 2008.

13 William Baumol and William Bowen, *Performing Arts, The Economic Dilemma: A Study of Problems Common to Theater, Opera, Music and Dance*, MIT Press, Cambridge, Massachusetts, 1968.

14 See Huon Fraser, 'The Labour Income Share in New Zealand: An Update', New Zealand Productivity Commission Staff Research Note 2018/1, 2018, p.7, www.productivity.govt.nz/sites/default/files/The%20Labour%20Income%20Share%20in%20New%20Zealand%20March%202018_0.pdf (accessed 3 February 2019).

15 Due to the rise in ticket prices implied this phenomenon was called Baumol's cost disease. See William J. Baumol with David Ferranti et al., *The Cost Disease: Why Computers Get Cheaper and Health Care Doesn't*, Yale University Press, New Haven, 2012.

16 We came across a version of the Baumol effect in the last chapter in observing that over time we spend a greater share of our income in sectors that are less productive.

17 Adapted from Lee, 'The productivity paradox'.

18 Lawrence Mishel and Josh Bivens,

'The zombie robot argument lurches on', Economic Policy Institute, 24 May 2017, www.epi.org/files/pdf/126750.pdf (accessed 15 December 2018).

19 There is a broader discussion about whether it is justifiable that people are paid according to their marginal productivity. In many cases jobs like teaching, nursing, or care work deliver high welfare benefits to others and to society. There is a strong case they should be paid more on this basis than on their productivity or simply with whatever result we get from the Baumol wage effect. See Turner, 'Capitalism in the age of robots', p.31.

20 For a discussion of how Baumol's wage effect could keep the economy on a relatively balanced growth path even with rapid automation, see Philippe Aghion, Benjamin F. Jones and Charles I. Jones, 'Artificial Intelligence and Economic Growth', NBER Working Paper 23928, 2017.

21 More unequal countries have even less social mobility. See OECD, *A Broken Social Elevator? How to Promote Social Mobility*, OECD Publishing, Paris, 2018, p.16, https://doi.org/10.1787/9789264301085-en (accessed 3 February 2019).

22 Kirsty Johnson, 'Big read: The gap between the rich and poor at university', *NZ Herald*, 15 September 2018, www.nzherald.co.nz/nz/news/article.cfm?c_id=1&objectid=12122733 (accessed 31 January 2018). The same holds even for just getting to university. In 2017 just 29 per cent of students from schools in deciles 1 to 3 achieved University Entrance, compared with 65.7 per cent of students in decile 8 to 10 schools. See 'Annual Report on NCEA and New Zealand Scholarship Data and Statistics (2017)', NZQA, p.16, www.nzqa.govt.nz/assets/About-us/Publications/stats-reports/ncea-annual-report-2017.pdf (accessed 3 February 2019).

23 Of course low productivity growth in general is a big problem, as discussed in Chapter Three. But, to the extent that there are high-productivity-growth sectors, demand for workers in them seems to be outstripping the supply of skilled workers available.

24 'Solving the Talent Shortage Build, Buy, Borrow and Bridge', Manpower Group New Zealand, 2018, https://downloads.manpowergroup.co.nz/hubfs/MPG_TalentShortageSurvey_Files/assets/MG_TalentShortage_2018_NZ_2.pdf?hsCtaTracking=a4dba5ee-5029-433e-a9ac-8cd7d990b89e%7C53a7f32b-883f-457d-9a37-f9db139936cf (accessed 15 December 2018).

25 The US manufacturing sector saw a big decline in employment in the global financial crisis but employment has been gradually rising since 2010. 'American factories could prosper if they find enough skilled workers', *The Economist*, 12 October 2017, www.economist.com/business/2017/10/12/american-factories-could-prosper-if-they-find-enough-skilled-workers (accessed 15 December 2018); 'Employment, Hours, and Earnings from the Current Employment Statistics survey (National)', United States Bureau of Labor Statistics, https://data.bls.gov/timeseries/CES3000000001 (accessed 15 December 2018).

26 Digital Skills Forum, *Digital Skills for a Digital Nation*, 2017, p.9, https://www.digitalskillsforum.nz/wp-content/uploads/2018/01/digital-skills-for-a-digital-nation-online.pdf (accessed 11 March 2019).

27 Thomas Edsall, 'Establishment Populism Rising', *New York Times*, 4 March 2015, www.nytimes.com/2015/03/04/opinion/establishment-populism-rising.html (accessed 18 December 2018).

28 Author interview with Daron Acemoglu, 24 April 2018.

29 As Chapter Four outlined, jobs can also come about through the productivity effect and increased demand, but generating new tasks

and new jobs remains fundamental in the long run.

30 Peter S. Goodman, 'The Robots Are Coming, and Sweden Is Fine', *New York Times*, 27 December 2017, www.nytimes.com/2017/12/27/business/the-robots-are-coming-and-sweden-is-fine.html (accessed 6 February 2019).

31 *Attitudes Towards the Impact of Digitisation and Automation on Daily Life*, European Commission, 2017, p.59, http://ec.europa.eu/commfrontoffice/publicopinion/index.cfm/Survey/getSurveyDetail/instruments/SPECIAL/surveyKy/2160 (accessed 18 December 2018).

32 'Child poverty statistics: Year end June 2018', Statistics New Zealand, 1 April 2019, https://www.stats.govt.nz/information-releases/child-poverty-statistics-year-ended-june-2018 (accessed 4 April 2019).

33 See M. Duncanson, G. Oben, A. Wicken, S. Morris, M.A. McGee and J. Simpson, *Child Poverty Monitor: Technical Report 2017*, New Zealand Child and Youth Epidemiology Service, 2017, p.63, http://hdl.handle.net/10523/7775 (accessed 15 January 2019).

34 Gerard Cotterell, Susan St John, Claire Dale and Yun So, 'Further fraying of the welfare safety net', Child Poverty Action Group, 2017, www.cpag.org.nz/assets/171208%20CPAG%20further%20fraying%20of%20the%20welfare%20safety%20WEB.pdf (accessed 15 December 2018).

35 Jonathan Boston, 'Reforming the welfare state', Stuff, 20 February 2018, www.stuff.co.nz/dominion-post/comment/101403547/reforming-the-welfare-state (accessed 15 December 2018).

36 Some benefits, like the jobseekers' allowance, are linked to inflation, the Consumer Price Index, but despite this there has been a major erosion compared to wages. Moreover, the accommodation supplement is not indexed to the CPI. For details of the erosion, see David Carey, 'Making New Zealand's economic growth more inclusive', OECD Economics Department Working Paper 1256, OECD Publishing, Paris, 2015, p.23, www.oecd.org/officialdocuments/publicdisplaydocumentpdf/?cote=ECO/WKP(2015)74&docLanguage=En (accessed 18 December 2018).

37 Accommodation supplements top up only a little – $70 in some areas – and even with the absolute maximum of $165 in New Zealand's most expensive areas that still doesn't reach the superannuation total. See 'Benefit rates at 1 April 2018', Ministry of Social Development, www.workandincome.govt.nz/products/benefit-rates/benefit-rates-april-2018.html#null (accessed 15 December 2018).

38 Bryce Wilkinson, 'Welfare, work and wellbeing: From benefits to better lives', The New Zealand Initiative, 28 November 2017, https://nzinitiative.org.nz/reports-and-media/reports/welfare-work-and-wellbeing-from-benefits-to-better-lives (accessed 15 December 2018).

39 Jacques Bughin, Eric Hazan, Susan Lund, Peter Dahlström, Anna Wiesinger and Amresh Subramaniam, *Skill Shift – Automation and the future of the workforce*, McKinsey Global Institute, 2018, p.13, www.mckinsey.com/~/media/McKinsey/Featured%20Insights/Future%20of%20Organizations/Skill%20shift%20Automation%20and%20the%20future%20of%20the%20workforce/MGI-Skill-Shift-Automation-and-future-of-the-workforce-May-2018.ashx (accessed 18 December 2018).

40 As the World Bank WDR 2019 highlights, a study across a range of rich countries found that a one-standard-deviation increase in complex problem-solving skills is associated with a 10 to 20 per cent higher wage. See World Bank, *World Development Report 2019: The*

Changing Nature of Work, World Bank, Washington, DC, 2019, p.72.

41 Bughin et al., *Skill Shift*, p.11; World Bank, *World Development Report 2019*, p.50.

42 *Worldwide Educating for the Future Index*, Economist Intelligence Unit, 2017, https://dkflato8y5dsg.cloudfront.net/uploads/5/80/eiu-yidan-prize-educating-for-the-future-wp-final.pdf (accessed 18 December 2018).

43 David Carey, 'Adapting to the changing labour market in New Zealand', OECD Economics Department Working Paper 1420, OECD Publishing, Paris, 2017, pp.23–24.

44 OECD, *Low-Performing Students: Why They Fall Behind and How to Help Them Succeed*, OECD Publishing, Paris, 2016, p.66.

45 Kirsty Johnston, 'Search continues for schools silver bullet', *NZ Herald*, 5 November 2015, www.nzherald.co.nz/nz/news/article.cfm?c_id=1&objectid=11540372 (accessed 15 December 2018). However, analysis from the PPTA suggests that even students that are not performing so well relative to their peers in New Zealand are doing okay compared to low-performing students in other OECD countries. See Liz Gordon, *Who achieves what in secondary schooling? A conceptual and empirical analysis*, PPTA, p.3, www.ppta.org.nz/dmsdocument/81 (accessed 18 December 2018); 'NZ among worst ranked for inequality in education – report', Radio New Zealand, 30 October 2018, www.radionz.co.nz/news/national/369779/nz-among-worst-ranked-for-inequality-in-education-report (accessed 15 December 2018).

46 Yekaterina Chzhen, Anna Gromada, Gwyther Rees, Jose Cuesta and Zlata Bruckauf, 'An Unfair Start: Inequality in Children's Education in Rich Countries', Innocenti Report Card 15, UNICEF Office of Research, Innocenti, Florence, 2018, p.17.

47 World Bank, *World Development Report 2019*, p.73.

48 See James Heckman, 'Skill formation and the economics of investing in disadvantaged children', *Science*, 312, 5782 (2006), pp.1900–2.

49 In addition to tertiary education's value for the individual and the job market, increased time in education for many people would mean that the portion of life for which people will work may shrink, making it easier to generate enough employment for all.

50 For a discussion of how a college education may not always be necessary in the future from an American perspective, see Ellen Ruppell Shell, *The Job: Work and Its Future in a Time of Radical Change*, Currency, New York, 2018.

51 *Profile & Trends 2016: Tertiary Education Outcomes and Qualification Completions*, Ministry of Education, Wellington, 2017, www.educationcounts.govt.nz/statistics/indicators/main/education-and-learning-outcomes/earnings_premiums_by_qualification (accessed 15 December 2018).

52 This refers to the OECD's estimates. See OECD, *Education at a Glance 2018: OECD Indicators*, OECD Publishing, Paris, 2018, p.98. The annual premium for people with bachelor's degrees compared to NCEA Level 1 to 3 is between 30 and 45 per cent depending on methodologies. In addition to the OECD, see Zaneta Park, Paul Mahoney, Warren Smart and Roger Smyth, 'Looking at the employment outcomes of tertiary education', Tertiary Education Occasional Paper 2013/02, Ministry of Education, 2013, www.educationcounts.govt.nz/publications/80898/looking-at-the-employment-outcomes-of-tertiary-education (accessed 15 December 2018).

53 'Population with tertiary education (indicator)', OECD, https://doi.org/10.1787/0b8f90e9-en (accessed 18 December 2018).

54 Carey, 'Adapting to the changing labour market in New Zealand', p.28.

55 Ibid., p.6.

56 Sylvia Dixon and David C. Maré, 'The Costs of Involuntary Job Loss: Impacts on Workers' Employment and Earnings', Motu Working Paper 13-03, Motu Economic and Public Policy Research, 2013, https://motu.nz/assets/Documents/our-work/population-and-labour/individual-and-group-outcomes/13-03.pdf (accessed 15 December 2018).

57 OECD, *Back to Work New Zealand: Improving the Re-Employment Prospects of Displaced Workers*, OECD Publishing, Paris, 2017, p.40.

58 Ibid., p.86.

59 Dani McDonald, 'Weta Digital running out of homes for award winning team', Stuff, 25 February 2018, www.stuff.co.nz/entertainment/film/101507890/weta-digital-running-out-of-homes-for-award-winning-team (accessed 15 December 2018).

60 See Edward Glaeser and David Maré, 'Cities and Skills', *Journal of Labor Economics*, 19, 2 (2001), pp.316–42; Edward Glaeser and Matthew Ressenger, 'The Complementarity between Cities and Skills', *Journal of Regional Science*, 50, 1 (2010), pp.221–44.

61 Edward Glaeser, *Triumph of the City: How Our Greatest Invention Makes Us Richer, Smarter, Greener, Healthier, and Happier*, Penguin Press, New York, 2011.

62 'The Economist house-price index', *The Economist*, https://infographics.economist.com/2017/HPI/index.html (accessed 15 December 2018); 'Global Housing Watch, Q2 2017', IMF, www.imf.org/external/research/housing/report/pdf/Q2_2017.pdf (accessed 15 December 2018).

63 'Full Ranking', TomTom Traffic Index, www.tomtom.com/en_gb/trafficindex/list?citySize=LARGE&continent=ALL&country=ALL (accessed 18 December 2018).

64 Dave Maré, 'Urban Productivity Estimation with Heterogeneous Prices and Labour', Motu Working Paper 16-21, http://motu-www.motu.org.nz/wpapers/16_21.pdf (accessed 18 December 2018).

65 Author interview with Patrick McVeigh, 22 May 2018.

66 For one perspective on how trade policy could help support employment growth despite rapid automation, see Michael Lind, 'Protect Yourself Before You Wreck Yourself', *Foreign Policy*, 16 July 2018, https://foreignpolicy.com/2018/07/16/protect-yourself-before-you-wreck-yourself-automation-manufacturing-jobs-technology (accessed 15 December 2018).

67 See 'Where did New Zealand export to in 2016', Atlas of Economic Complexity, http://atlas.cid.harvard.edu/explore/?country=166&partner=undefined&product=undefined&productClass=HS&startYear=undefined&target=Partner&year=2016 (accessed 18 December 2018).

68 Author interview with Larry Summers, 18 October 2018.

69 Geoff Bascand, 'Changing Dynamics in Household Behaviour: What do they mean for inflationary pressures?', Reserve Bank of New Zealand, 2016, www.rbnz.govt.nz/research-and-publications/speeches/2016/speech2016-11-22 (accessed 15 December 2018).

70 Weak decoupling is also known as relative decoupling, whilst strong decoupling is also known as absolute decoupling.

71 This is over the 1995 to 2011 period. See OECD, *Green Growth Indicators 2017*, OECD Green Growth Studies, OECD Publishing, Paris, 2017, p.35.

72 Ibid. This focuses only on carbon-dioxide emissions, which understates how significant New Zealand's emissions are, because methane, primarily from agriculture, is a significant part of our emissions (even

considering much of our agricultural produce is exported). Given the fact agriculture has been exempted from our emissions trading scheme and has been globally a challenging sector in which to reduce emissions, New Zealand's efforts at decoupling all greenhouse gases are likely to be worse than the figures presented here, as this captures all our GDP growth but not all of our emissions.

73 OECD, *Green Growth Indicators 2017*, p.35.

74 For a striking recent example of this, see Robin Harding, 'Cancer or climate change: which cure would you invest in?', *Financial Times*, 27 November 2018, www.ft.com/content/3a8a6410-f230-11e8-ae55-df4bf40f9d0d (accessed 18 December 2018).

75 Thomas O. Wiedmann, Heinz Schandl, Manfred Lenzen, Daniel Moran, Sangwon Suh, James West and Keiichiro Kanemoto, 'The material footprint of nations', *Proceedings of the National Academy of Sciences*, 112, 20 (May 2015), pp.6271–76. There is still of course robust debate about these findings, with Chris Goodall for example arguing that the United Kingdom has peaked in its material footprint. See Chris Goodall, '"Peak Stuff": Did the UK reach a maximum use of material resources in the early part of the last decade?', 31 October 2011, http://static.squarespace.com/static/545e40d0e4b054a6f8622bc9/t/54720c6ae4b06f326a850 2f9/1416760426697/Peak_Stuff_17.10.11.pdf (accessed 18 December 2018).

76 Johan Rockström, Will Steffen, Kevin Noone, Åsa Persson, Stuart Chapin III, Eric Lambin, Timothy Lenton et al., 'A safe operating space for humanity', *Nature*, 461 (2009), pp.472–75.

77 Kate Raworth, *Doughnut Economics: Seven Ways to Think Like a 21st Century Economist*, Chelsea Green Publishing, Vermont, 2017, loc. 3192 of 4507, Kindle.

Chapter Six

1 Erik Brynjolfsson and Andrew McAfee, *The Second Machine Age: Work, Progress, and Prosperity in a Time of Brilliant Technologies*, W.W. Norton & Company, New York, 2014; Martin Ford, *Rise of the Robots: Technology and the Threat of a Jobless Future*, Basic Books, New York, 2015. There are numerous others, including Paul Mason, the former economics editor for the BBC's flagship current-affairs programme *Newsnight*, who have called for a UBI and radical reduction of the working day. See Paul Mason, 'Automation may mean a post-work society but we shouldn't be afraid', *Guardian*, 17 February 2016, www.theguardian.com/sustainable-business/2016/feb/17/automation-may-mean-a-post-work-society-but-we-shouldnt-be-afraid (accessed 18 December 2018).

2 Richard Freeman, 'Who owns the robots rules the world', *IZA World of Labor*, 5 (2015), https://doi.org/10.15185/izawol.5 (accessed 20 February 2019). New Zealand's Robert Wade has proposed a similar idea, although not explicitly to deal with automation. See Robert H. Wade, 'Get profit share to support a more equal income distribution', *Financial Times*, 20 August 2015, www.ft.com/content/5c1b8994-4732-11e5-af2f-4d6e0e5eda22 (accessed 17 December 2018). Another New Zealander, Keith Rankin, an economist at Unitec, has argued for public equity dividends as well, although again not with the explicit aim of addressing automation; see Keith Rankin, 'From Universal Basic Income to Public Equity Dividends', Briefing Papers, 6 March 2018, http://briefingpapers.co.nz/from-universal-basic-income-to-public-equity-dividends (accessed 17 December 2018).

3 Studies of trials of negative income tax in the 1970s found that people did choose to work less. See Jodie T. Allen, 'Negative income tax', *The*

Concise Encyclopedia of Economics, Library of Economics and Liberty, www.econlib.org/library/Encl/NegativeIncomeTax.html (accessed 17 December 2018).

4 Changes to societal expectations and norms around employment and work are themselves likely to have a significant effect on individual decisions about whether and how much to work. Moreover, with a UBI, gift shares and higher taxes people are considerably more likely to choose not to work, as the marginal benefit of doing so would not be large enough to compensate for the lost leisure time.

5 See David Graeber, *Bullshit Jobs: A Theory*, Simon and Schuster, New York, 2018, pp.9–10.

6 Andrew Anthony, 'Bullshit Jobs: A Theory review – laboured rant about the world of work', *Guardian*, 27 May 2018, www.theguardian.com/books/2018/may/27/bullshit-jobs-a-theory-david-graeber-review-laboured-rant (accessed 17 December 2018).

7 See Tara Westover, *Educated: A Memoir*, Random House, New York, 2018, for a great discussion of this.

8 Derek Thompson, 'A world without work', *The Atlantic*, July/August 2015, www.theatlantic.com/magazine/archive/2015/07/world-without-work/395294 (accessed 17 December 2018).

9 Bill Davies, Joe Dromey, Clare McNeil, Charlotte Snelling and Craig Thorley, 'Working well: A plan to reduce long term sickness absence', The Progressive Policy Think Tank, 15 February 2017, www.ippr.org/publications/working-well-a-plan-to-reduce-long-term-sickness-absence (accessed 17 December 2018); Simon Collins, 'Winz moves early to bring in tough stance on benefits', *NZ Herald*, 6 April 2010, www.nzherald.co.nz/nz/news/article.cfm?c_id=1&objectid=10636491 (accessed 17 December 2018).

10 This essay was presented in 1930 as a lecture in Madrid and later that year in two installments in the *Nation and Athenaeum*. See John Maynard Keynes, 'Economic Possibilities for Our Grandchildren', in *Essays in Persuasion*, Palgrave Macmillan, London, 2010.

11 See 'Average usual weekly hours worked on the main job (time series)', OECD.Stat, https://stats.oecd.org/Index.aspx?DataSetCode=AVE_HRS# (accessed 15 February 2019).

12 We have declined from 1,834 hours per year in 1986 to 1,753 hours per year in 2017. In addition, since 2009 the decline in working hours has flattened out and the hours have even increased a little. See 'Hours worked (indicator)', OECD, 2018, https://doi.org/10.1787/47be1c78-en (accessed 14 December 2018).

13 Tim Harford, 'Would basic incomes or basic jobs be better when robots take over?', *Financial Times*, 29 June 2018, www.ft.com/content/c1c8a28e-7a17-11e8-8e67-1e1a0846c475 (accessed 17 December 2018).

14 In his own discussion of the future of work, Jim Arrowsmith of the Massey University School of Management raises the possibility that we should rethink our obsession with growth, and notes that many wealthy countries rank as less happy than places like Costa Rica and Bhutan. See Jim Arrowsmith, 'AI, Robotics and Automation – Implications for work, employment and society', *Big Issues in Business*, October 2017, www.massey.ac.nz/shadomx/apps/fms/fmsdownload.cfm?file_uuid=2BEDD3C2-3EA8-4817-8492-EA6F36A2442C (accessed 18 December 2018).

15 See the discussion in Tim Jackson, *Prosperity Without Growth*, Earthscan, Abingdon, 2009, pp.98–102.

16 Philip Brickman, Dan Coates and Ronnie Janoff-Bulman, 'Lottery winners and accident victims: Is happiness relative?', *Journal of*

Personality and Social Psychology, 36, 8 (1978), p.917.

17 See Daniel Kahneman and Angus Deaton, 'High income improves evaluation of life but not emotional well-being', *Proceedings of the National Academy of Sciences*, 107, 38 (2010), pp.16,489–93. It is worth noting that many Americans, and many New Zealanders, still earn substantially less than US$75,000.

18 The question of whether there is a satiation point for life satisfaction is not entirely settled. The view that there is no satiation is captured both by Kahneman and Deaton as well as by Betsey Stevenson and Justin Wolfers, 'Subjective Well-Being and Income: Is There Any Evidence of Satiation?', *American Economic Review*, 103, 3 (2013), pp.598–604. For the view that life satisfaction may actually stop increasing altogether at some point regardless of income, see Andrew T. Jebb, Louis Tay, Ed Diener and Shigehiro Oishi, 'Happiness, income satiation and turning points around the world', *Nature Human Behaviour*, 2 (2018), pp.33–38. They suggest that in Australia and New Zealand income's impact on life evaluation reaches a satiation point at US$125,000 per individual; those with kids have a higher point. Notably, this was the highest satiation point for any region they studied and is a very long way above actual median income in New Zealand. However, as even the authors acknowledge, this paper is one of the few studies that finds not just that it takes more extra income to improve life satisfaction the richer you are, but that actually extra income may not improve life satisfaction at all.

19 On expectations see Lara B. Aknin, Michael I. Norton and Elizabeth W. Dunn, 'From wealth to well-being? Money matters, but less than people think', *Journal of Positive Psychology*, 4, 6 (2009), pp.523–27. On needing more money for the same increase in happiness most studies find that the relationship is log-linear, which means that if you double income for a poor person or a rich person, the impact on life satisfaction is the same, but of course doubling is a far larger amount of money for the rich person than the poor person. The rich therefore require more money in absolute terms than the poor to see the same improvement in life satisfaction. See, for example, Stevenson and Wolfers, 'Subjective Well-Being and Income'.

20 Ashley V. Whillans, Elizabeth W. Dunn, Paul Smeets, Rene Bekkers and Michael I. Norton, 'Buying time promotes happiness', *Proceedings of the National Academy of Sciences*, 114, 32 (July 2017), pp.8523–27, https://doi.org/10.1073/pnas.1706541114 (accessed 16 January 2019).

21 Ed Diener, Ronald Inglehart and Louis Tay, 'Theory and validity of life satisfaction scales', *Social Indicators Research*, 112, 3 (2013), pp.112–497, https://doi.org/10.1007/s11205-012-0076-y (accessed 16 January 2019); Tim Kasser, *The High Price of Materialism*, MIT Press, Cambridge, Massachusetts, 2003.

22 This could be happening in a variety of ways, such as making us jealous or resentful, making us feel ashamed or humiliated, or simply meaning we focus on income to the detriment of other things that make us happier.

23 Moreover, a world without ever-increasing consumption on aggregate does not mean people cannot consume new things. It just means that the sum total of their consumption is not more than before; what they consume could still change dramatically to follow changing tastes and preferences.

24 Kate Raworth, *Doughnut Economics: Seven Ways to Think Like a 21st Century Economist*, Chelsea Green Publishing, Vermont, 2017, loc. 3943 of 4507, Kindle. See also Jackson, *Prosperity Without Growth*.

25 Just as for workers deciding whether to work or not in Rachel and Wei's

world, whether businesses aim for constant growth in profits or just a sustained level of profits each year will also depend on the extent to which their additional income would be taxed and on their level of satisfaction with their existing income from a basic income, gift stocks and (for businesses) their existing revenues.

26 Robert Skidelsky and Edward Skidelsky, *How Much Is Enough? Money and the Good Life*, Other Press, New York, 2012.

27 These paragraphs draw on Susan Bidwell and Sarah Colhoun, 'Health impact and the public health response to major job losses in small communities – An overview of the international and New Zealand literature', Canterbury District Health Board, 2 May 2013, www.cph.co.nz/wp-content/uploads/healthimpactjobloss.pdf (accessed 17 December 2018).

28 V. Keefe, P. Reid, C. Ormsby, B. Robson, G. Purdie, J. Baxter and Ngāti Kahungunu Iwi Incorporated, 'Serious health events following involuntary job loss in New Zealand meat processing workers', *International Journal of Epidemiology*, 31, 6 (2002), pp.1155–61.

29 Richard E. Lucas et al., 'Unemployment Alters the Set Point for Life Satisfaction', *Psychological Science*, 15, 1 (January 2004), pp.8–13.

30 See 'Unemployment, (re) employment and well-being', WhatWorksWellbeing, March 2017, https://whatworkswellbeing.files.wordpress.com/2017/02/unemployment-reemployment-wellbeing-briefing-march-2017.pdf (accessed 17 December 2018). See also M. Luhmann, W. Hofmann, M. Eid and R.E. Lucas, 'Subjective well-being and adaptation to life events: A meta-analysis', *Journal of Personality and Social Psychology*, 102, 3 (2012), pp.592–615. https://doi.org/10.1037/a0025948 (accessed 16 January 2019).

31 See Jenny Marlar, 'The Emotional Cost of Underemployment', *Gallup*, 9 March 2010, https://news.gallup.com/poll/126518/emotional-cost-underemployment.aspx (accessed 17 December 2018). See also A. Milner and A.D. LaMontagne, 'Underemployment and mental health: Comparing fixed-effects and random-effects regression approaches in an Australian working population cohort', *Occupational and Environmental Medicine*, 74 (2017), pp.344–50; and Maureen F. Dollard and Anthony H. Winefield, 'Mental health: Overemployment, underemployment, unemployment and healthy jobs', *Australian e-Journal for the Advancement of Mental Health*, 1, 3 (2002), pp.170–95, https://doi.org/10.5172/jamh.1.3.170 (accessed 16 January 2019).

32 Pedro Nicolaci da Costa, 'Robots are going to take a lot of jobs – here's what we could do about it', *Business Insider*, 30 April 2017, www.businessinsider.com/policy-responses-to-automation-and-robots-taking-jobs-2017-4 (accessed 17 December 2018).

33 Mihaly Csikszentmihalyi and Judith LeFevre, 'Optimal experience in work and leisure', *Journal of Personality and Social Psychology*, 56, 5 (1989), pp.815–22.

34 Mihaly Csikszentmihalyi, *Flow: The Psychology of Optimal Experience*, Harper and Row, New York, 1990, p.69.

35 Csikszentmihalyi and LeFevre, 'Optimal experience in work and leisure'.

36 Ibid.

37 As Csikszentmihalyi and LeFevre put it, we have a paradox of people having 'many more positive feelings at work than in leisure, yet saying that they "wish to be doing something else" when they are at work, not when they are in leisure'. Ibid., p.821.

38 Brian Groom, 'Self-employed happier in work than employees, UK survey finds', *Financial Times*, 27 May 2014,

www.ft.com/content/93e651ba-e5bd-11e3-aeef-00144feabdc0 (accessed 17 December 2018); Peter Warr and Ilke Inceoglu, 'Work Orientations, Well-Being and Job Content of Self-Employed and Employed Professionals', *Work, Employment and Society*, 32, 2 (2018), pp.292–311.

39 'Time Use Survey 2009–10', Stats NZ, http://archive.stats.govt.nz/browse_for_stats/people_and_communities/time_use/TimeUseSurvey_HOTP2009-10/Data%20Quality.aspx (accessed 18 December 2018).

40 Jinmoo Heo, Youngkhill Lee, Paul M. Pedersen and Bryan P. McCormick, 'Flow Experience in the Daily Lives of Older Adults: An Analysis of the Interaction between Flow, Individual Differences, Serious Leisure, Location, and Social Context', *Canadian Journal on Aging/La Revue Canadienne du Vieillissement*, 29, 3 (2010), pp.411–23, www.calvin.edu/~yl33/documents/HeoCJAarticle.pdf (accessed 16 January 2019).

41 Walter Scheidel, *The Great Leveler: Violence and the History of Inequality from the Stone Age to the Twenty-First Century*, Princeton University Press, Princeton, 2018.

42 See, for example, Duncan Green, 'What do we know about the politics of reducing inequality? Not much – it's time to find out', *From Poverty to Power* (blog), Oxfam, 13 February 2015, https://oxfamblogs.org/fp2p/what-do-we-know-about-the-politics-of-reducing-inequality-not-much-its-time-to-find-out (accessed 18 December 2018). For more recent examples of changes, including declines, in inequality since 1990, see Joe Hassell, 'Is income inequality rising around the world?', Our World in Data, 19 November 2018, https://ourworldindata.org/income-inequality-since-1990 https://ourworldindata.org/income-inequality-since-1990 (accessed 15 February 2019).

43 Piketty shows in recent history *r*, the returns to capital, have been greater than *g*, economic growth, meaning that those who already have capital get wealthier faster than those who earn primarily through wages, which are linked to economic growth. See Thomas Piketty, *Capital in the Twenty-First Century*, Harvard University Press, Cambridge, Massachusetts, 2013.

44 Nicholas Carr, *The Glass Cage: How Our Computers Are Changing Us*, W.W. Norton & Company, New York, 2015, p.227.

45 This section draws from Peter Skilling, 'Egalitarian myths in New Zealand: A review of public opinion data on inequality and redistribution', *New Zealand Sociology*, 28, 2 (2013), p.22.

46 Karl Ove Moene and Michael Wallerstein, 'Inequality social insurance and redistribution', *American Political Science Review*, 95, 4 (2001), pp.859–74.

47 Jon Walker, 'Machine Learning in Manufacturing – Present and Future Use-Cases', EMERJ, 5 December 2018, www.techemergence.com/machine-learning-in-manufacturing (accessed 17 December 2018).

48 Chris Baraniuk, 'How talking machines are taking call centre jobs', BBC, 14 August 2018, www.bbc.com/news/business-45272835 (accessed 17 December 2018).

49 Martin Ford (*Rise of the Robots*) claims that technology will encroach into tasks that have typically been more human, but this implies we cannot find new things we humans might want to do or problems we might want to solve. Yet as Nick Hannauer, the American entrepreneur, puts it, 'prosperity in human societies is best understood as the accumulation of solutions to human problems. We won't run out of work until we run out of problems.' That still looks incredibly far off. Quoted in Tim O'Reilly, *WTF? What's the Future and Why It's Up to Us*, HarperBusiness, New York, 2017, p.300.

50 James Manyika, Susan Lund, Michael Chui, Jacques Bughin, Jonathan Woetzel, Parul Batra, Ryan Ko and Saurabh Sanghv, *Jobs Lost, Jobs Gained: Workforce Transitions in a Time of Automation*, McKinsey Global Institute, 2017, p.50, www.mckinsey.com/~/media/mckinsey/featured%20insights/future%20of%20organizations/what%20the%20future%20of%20work%20will%20mean%20for%20jobs%20skills%20and%20wages/mgi-jobs-lost-jobs-gained-report-december-6-2017.ashx (accessed 30 January 2019).

51 See, for example, the impact of computer spreadsheets discussed in Greg Ip, 'We Survived Spreadsheets, and We'll Survive AI', *Wall Street Journal*, 2 August 2017, www.wsj.com/articles/wesurvived-spreadsheets-and-well-survive-ai-1501688765 (accessed 2 January 2019).

52 Manyika et al., *Jobs Lost, Jobs Gained*, p.4.

53 See Ryan Avent, *The Wealth of Humans, Work, Power, and Status in the Twenty-first Century*, Macmillan, New York, 2016, loc. 914 of 3883, Kindle.

54 Max Nisen, 'The most popular programming languages are rapidly changing', *QUARTZ*, 8 April 2015, https://qz.com/378939/the-most-popular-programming-languages-are-rapidly-changing (accessed 17 December 2018); Kevin Montrose, 'The Changing Landscape of Programming Technologies', Stack Overflow Business, 25 January 2017, www.stackoverflowbusiness.com/blog/the-changing-landscape-of-programming-technologies (accessed 17 December 2018).

55 James Bessen, *Learning by Doing: The Real Connection between Innovation, Wages, and Wealth*, Yale University Press, New Haven, 2015, p.63.

56 In New Zealand the company Perpetual Guardian trialled and then adopted a four-day work week, in which it also kept pay at the same level it had been for a five-day week.

57 Even without changes to redistribution some people are already choosing to save as much as possible and retire radically earlier. See Chris Stokel-Walker, 'FIRE: The movement to live frugally and retire decades early', BBC, 2 November 2018, www.bbc.com/capital/story/20181101-fire-the-movement-to-live-frugally-and-retire-decades-early (accessed 15 February 2019).

58 Brigid Delaney, 'Underworked, overworked: Australia's two-speed workforce', *Guardian*, 30 May 2013, www.theguardian.com/world/2013/may/31/underworked-overworked-two-speed-workforce (accessed 17 December 2018); and Adam Bandt, 'Too many people are overworked or underworked', *Sydney Morning Herald*, 9 March 2015, www.smh.com.au/opinion/too-many-people-are-overworked-or-underworked-20150308-13y57x.html (accessed 17 December 2018).

Chapter Seven

1 For further details see Ha-Joon Chang, *Bad Samaritans: The Myth of Free Trade and the Secret History of Capitalism*, Bloomsbury Press, London, 2009, p.3.

2 Jennifer Gidley, *The Future: A Very Short Introduction*, Oxford University Press, 2017, p.2.

3 Ibid., p.137.

4 Tim O'Reilly, *WTF? What's the Future and Why It's Up to Us*, HarperBusiness, New York, 2017.

5 Georg Graetz and Guy Michaels, 'Robots at Work', *Review of Economics and Statistics*, 100, 5 (December 2018), www.mitpressjournals.org/doi/abs/10.1162/rest_a_00754 (accessed 18 December 2018).

6 Author interview with Larry Summers, 18 October 2018.

7 For a discussion and affirmation of the power of socio-political choices to shape the future of work, see Göran Roos and Zara Shroff, 'What will happen to the jobs? Technology-enabled productivity improvement –

good for some, bad for others', *Labour & Industry: A Journal of the Social and Economic Relations of Work*, 27, 3 (2017), pp.165–92.

8 Joseph Stiglitz also includes a similar approach to this as a key policy recommendation for addressing the future of work. See Joseph Stiglitz, 'The coming great transformation', *Journal of Policy Modeling*, 39, 4 (2017), pp.625–38.

9 Federico Cingano, 'Trends in Income Inequality and Its Impact on Economic Growth', OECD Social, Employment and Migration Working Paper 163, OECD Publishing, Paris, 2014, https://doi.org/10.1787/5jxrjncwxv6j-en (accessed 18 December 2018). For one critique, see Eric Crampton, 'OECD on inequality', Offsetting Behaviour, 11 December 2014, http://offsettingbehaviour.blogspot.com/2014/12/oecd-on-inequality.html (accessed 15 February 2019).

10 See 'Social spending (indicator)', OECD, 2019, https://data.oecd.org/socialexp/social-spending.htm (accessed 16 February 2019).

11 'Does income inequality hurt economic growth?', OECD, December 2014, www.oecd.org/els/soc/Focus-Inequality-and-Growth-2014.pdf (accessed 18 December 2018); Jonathan Ostry, Andrew Berg and Charalambos Tsangarides, 'Redistribution, Inequality, and Growth', IMF Staff Discussion Note 14/02, April 2014, www.imf.org/external/pubs/ft/sdn/2014/sdn1402.pdf (accessed 16 February 2019). Both are clear that taken to extremes this can be destructive, but very few examples of what governments have done fall into the extreme category.

12 See OECD, *Back to Work: New Zealand: Improving the Re-employment Prospects of Displaced Workers*, OECD Publishing, Paris, 2017.

13 The cost of this may be a challenge for some small businesses, although in a world in which technological progress has continued and productivity has increased this problem should be smaller than it is today. If firms can earn more with less, the effect of productivity growth, redistribution measures like this would become easier. Moreover, even in Denmark, which is well known for its flexicurity model, which includes highly flexible labour markets, the notice period for layoffs for individuals is three months.

14 See Kate Evans, 'Daddy's home! Why aren't more men taking paternity leave?', Noted, 27 June 2018, www.noted.co.nz/currently/paternity-leave-new-zealand-fathers-taking-paid-leave (accessed 18 December 2018); and 'Why Swedish men take so much paternity leave', *The Economist*, 23 July 2014, www.economist.com/the-economist-explains/2014/07/22/why-swedish-men-take-so-much-paternity-leave (accessed 18 December 2018).

15 Max Rashbrooke, *Government for the Public Good: The Surprising Science of Large-Scale Collective Action*, Bridget Williams Books, Wellington, 2018, loc. 3896 of 9420, Kindle.

16 This is based specifically on the extent to which low socio-economic status increases the likelihood of poor performance in mathematics. OECD, *Low-Performing Students: Why They Fall Behind and How to Help Them Succeed*, OECD Publishing, Paris, 2016, p.66.

17 Tyler Cowen suggests teachers may become more akin to motivators or coaches; Tyler Cowen, *Average Is Over: Powering America beyond the Age of the Great Stagnation*, Dutton, New York, 2013.

18 *Worldwide Educating for the Future Index*, Economist Intelligence Unit, 2017, p.18, https://dkf1ato8y5dsg.cloudfront.net/uploads/5/80/eiu-yidan-prize-educating-for-the-future-wp-final.pdf (accessed 18 December 2017).

19 Jim Arrowsmith, 'AI, Robotics and Automation – Implications for work, employment and society',

Big Issues in Business, October 2017, p.22, www.massey.ac.nz/shadomx/apps/fms/fmsdownload.cfm?file_uuid=2BEDD3C2-3EA8-4817-8492-EA6F36A2442C (accessed 16 February 2019).

20 See David Autor, 'Will automation take away all our jobs?', TED, www.ted.com/talks/david_autor_why_are_there_still_so_many_jobs/transcript?language=en (accessed 20 February 2019).

21 Some may favour targeted scholarships for candidates from poorer backgrounds, given many tertiary students come from middle-class families and on average at least receive a pay premium upon graduation.

22 See 'What's the value of a tertiary degree anymore?', *Listener*, 6 April 2017, www.noted.co.nz/currently/education/whats-the-value-of-a-tertiary-degree-anymore (accessed 16 February 2019).

23 As Chapter Five showed, New Zealand has good re-employment rates, but workers tend to take a much larger hit to wages than workers in other countries do in their new job.

24 We spent 0.33 per cent of GDP in 2014, in the bottom third of OECD countries, and far below the leaders in the OECD, Denmark and Sweden, which spent 1.6 and 1.1 per cent respectively. See OECD, *Back to Work*.

25 Ibid., p.14.

26 Ibid., p.87.

27 David Carey, 'Adapting to the changing labour market in New Zealand', OECD Economics Department Working Paper 1420, OECD Publishing, Paris, 2017.

28 Kinley Salmon, 'Education and Training – Lessons from Denmark', The Future of Work in New Zealand, https://pdfs.semanticscholar.org/99da/9d730860306f384f9bc6e80018b4292ad0ef.pdf?_ga=2.34458593.1126681121.1531608847-540307580.1531608847 (accessed 18 December 2018). I wrote this report as a background paper for the New Zealand Labour Party's Future of Work Commission. I was not paid for this work.

29 Employers could also play a much more significant role in supporting workers to find new work. In Sweden 'job security councils' run by employers and unions help workers to search for a new job and provide intensive retraining as soon as the worker learns they will be laid off. Employers fund these councils in part because it means union resistance to layoffs is lower, which in turn allows the economy to run more smoothly. In both Sweden and Denmark re-employment rates are good while workers face much lower drops in pay in their new jobs than in New Zealand. See OECD, *Back to Work*.

30 See Edward Glaeser and David Maré, 'Cities and Skills', *Journal of Labor Economics*, 19, 2 (2001), pp.316–42; Dan Levy, Ricardo Hausmann, Miguel Santos, Luis Espinoza and Miguel Flores, 'Why is Chiapas Poor?', CID Working Paper 300, 1 March 2016, https://ssrn.com/abstract=2880665 (accessed 18 December 2018).

31 Dave Maré, 'Urban Productivity Estimation with Heterogeneous Prices and Labour', Motu Working Paper 16-21, http://motu-www.motu.org.nz/wpapers/16_21.pdf (accessed 18 December 2018).

32 *Prosperity in Auckland*, Auckland Tourism, Events and Economic Development, 2018, www.aucklandnz.com/sites/build_auckland/files/media-library/documents/Auckland-Prosperity-Index-Report-March-2018.pdf (accessed 3 December 2018). Low levels of prosperity also map to some degree onto ethnicity in Auckland. For example, a number of areas with large Pacific Island communities score relatively low on the prosperity index. For a broader discussion of Pacific communities in New Zealand see Damon Salesa, *Island Time: New Zealand's Pacific Futures*, Bridget

Williams Books, Wellington, 2017.

33 *New Zealand Property Focus: Holding the Key*, ANZ Research, July 2018, p.5, www.anz.co.nz/resources/4/a/4a28ef88-3f49-4c17-99e4-9549d39e0934/ANZ-PropertyFocus-20180718.pdf?MOD=AJPERES (accessed 4 December 2018).

34 We rank worst in the OECD for mismatches between our jobs and both qualification level and field of study. See OECD, *Skills Matter: Further Results from the Survey of Adult Skills*, OECD Publishing, Paris, 2016, p.133.

35 Carey, 'Adapting to the changing labour market in New Zealand'.

36 This appears to be part of the problem in the USA. See Greg Kaplan and Sam Schulhofer-Wohl, 'Understanding the long-run decline in interstate migration', *International Economic Review*, 58, 1 (2017), pp.57–94.

37 See Don Brash, quoted in Sam Sachdeva, 'Analysis: Can we have affordable houses without prices dropping sharply?', Stuff, 8 July 2016, www.stuff.co.nz/national/politics/81916481/analysis-can-we-have-affordable-houses-without-prices-dropping-sharply (accessed 10 December 2018).

38 See 'Reshaping Urban Mobility with Autonomous Vehicles Lessons from the City of Boston', World Economic Forum in collaboration with the Boston Consulting Group, June 2018, www.weforum.org/reports/reshaping-urban-mobility-with-autonomous-vehicles-lessons-from-the-city-of-boston (accessed 18 December 2018). In addition, if driverless cars are still primarily single-occupant they will almost certainly result in far more congestion than most public-transport options.

39 Our existing trade with Latin America is very low despite being as close to New Zealand as much of East Asia. Kinley Salmon, 'Kinley Salmon: Cure on hand for Chinese headache', *NZ Herald*, 3 February 2016, www.nzherald.co.nz/business/news/article.cfm?c_id=3&objectid=11583650 (accessed 18 December 2018).

40 See 'Southeast Asia Surging in CID's New Global Growth Projections', Center for International Development Harvard, http://atlas.cid.harvard.edu/rankings/growth-projections (accessed 18 December 2018).

41 'New Zealand–India FTA', New Zealand Foreign Affairs and Trade, www.mfat.govt.nz/en/trade/free-trade-agreements/agreements-under-negotiation/india (accessed 18 December 2018).

42 The inadequacy of support for those negatively affected by trade in the USA and the United Kingdom has been problematic and is increasingly viewed as behind many recent political changes. See 'America's programme to help trade's losers needs fixing', *The Economist*, 29 January 2017, www.economist.com/finance-and-economics/2017/06/29/americas-programme-to-help-trades-losers-needs-fixing (accessed 18 December 2018).

43 Author interview with Larry Summers.

44 Joseph Stiglitz also lists this as perhaps the most important policy for creating a good future of work. See Stiglitz, 'The coming great transformation'.

45 Author interview with Larry Summers.

46 Stiglitz, 'The coming great transformation'.

47 See Max Rashbrooke (ed.), *Inequality: A New Zealand Crisis*, Bridget Williams Books, Wellington, 2015; and Tony Atkinson, *Inequality: What can be done?*, Harvard University Press, Cambridge, Massachusetts, 2015.

48 On competition see OECD, *OECD Economic Surveys: New Zealand 2017*, OECD Publishing, Paris, 2017, p.34.

New Zealand's minimum wage for full-time workers is already relatively high compared to the median wage, at 60 per cent, compared with other countries in the OECD. See 'Minimum relative to average wages of full-time workers (indicator)', OECD.Stat, https://stats.oecd.org/Index.aspx?DataSetCode=MIN2AVE# (accessed 18 December 2018).

49 Stiglitz, 'The coming great transformation'.

50 One technical approach is to encourage widespread adoption of new technologies across more businesses. This can help reduce wage differences between businesses in the same sector by reducing major productivity differences between different firms. Yet wage negotiations, as the chapter discusses, are also important.

51 Business New Zealand leader Kirk Hope is concerned about whether wage compacts reduce their flexibility and the productivity of workers but is also engaged in discussions on how to make these work. See 'Employment law changes won't help business growth – CEO', Radio New Zealand, 11 September 2018, www.radionz.co.nz/news/business/366211/employment-law-changes-won-t-help-business-growth-ceo (accessed 18 December 2018). There is also the possibility that raising low wages a lot might hasten automation in the long run, although even if that is the case, as this book has described, with the right policies there may be ways to support people into better, higher-paid jobs.

52 'Table I.7. Top statutory personal income tax rate and top marginal tax rates for employees', OECD.Stat, https://stats.oecd.org/index.aspx?DataSetCode=TABLE_I7 (accessed 18 December 2018). While there is no capital gains tax, the extended bright line test means profits on investment properties sold within five years of when they are purchased can be taxed.

53 See Rashbrooke (ed.), *Inequality*, loc. 862 of 5060, Kindle.

54 Ostry et al., 'Redistribution, Inequality, and Growth'; Cingano, 'Trends in Income Inequality'.

55 See previous references as well as Roman Arjona, Maxime Ladaique and Mark Pearson, 'Growth, Inequality and Social Protection', OECD Labour Market and Social Policy Occasional Paper 51, OECD, 2001.

56 See OECD, *Effective Carbon Rates 2018: Pricing Carbon Emissions through Taxes and Emissions Trading*, OECD Publishing, Paris, 2018.

57 This is only to meet the 2°C target, which would still mean major impact from climate change. 'Low-emissions economy – Draft Report', New Zealand Productivity Commission, April 2018, www.productivity.govt.nz/sites/default/files/Productivity%20Commission_Low-emissions%20economy_Draft%20report.pdf (accessed 18 December 2018).

58 This approach is often called a fee and dividend model; see James Hansen, 'Environment and Development Challenges: The Imperative of a Carbon Fee and Dividend', Earth Institute, Columbia University, 11 November 2015, http://csas.ei.columbia.edu/2015/11/11/environment-and-development-challenges-the-imperative-of-a-carbon-fee-and-dividend (accessed 18 December 2018); and John Horowitz, Julie-Anne Cronin, Hannah Hawkins, Laura Konda and Alex Yuskavage, 'Methodology for Analyzing a Carbon Tax', Office of Tax Analysis Working Paper 115, January 2017, www.treasury.gov/resource-center/tax-policy/tax-analysis/Documents/WP-115.pdf (accessed 18 December 2018). This US Treasury analysis finds that 60 per cent of families would get an increase in after-tax income thanks to such a scheme.

59 The Green Investment Fund, announced in late 2018, is a first step in that direction. Government

investments can have a big impact on our energy mix; fracking was developed with substantial support from the American government. Its impacts on climate change though are mixed. Fracking makes it easier to access both natural gas and oil in some places. Natural gas is cleaner than oil but fracking provides shale oil too, and has also reduced the need for major renewable investments. See Michael Shellenberger and Ted Nordhaus, 'A boom in shale gas? Credit the feds', *Washington Post*, 16 December 2011, www.washingtonpost.com/opinions/a-boom-in-shale-gas-credit-the-feds/2011/12/07/gIQAecFIzO_story.html?utm_term=.354fe0147f06 (accessed 18 December 2018).

60 Daron Acemoglu, Ufuk Akcigit, Douglas Hanley and William Kerr, 'Transition to clean technology', *Journal of Political Economy*, 124, 1 (2016), pp.52–104.

61 Nicholas Carr, *The Glass Cage: How Our Computers Are Changing Us*, W.W. Norton & Company, New York, 2015.

62 This section draws on *Climate Change Implications for New Zealand*, Royal Society of New Zealand, 2016, https://royalsociety.org.nz/assets/Uploads/Climate-change-implications-for-NZ-2016-report-web2.pdf (accessed 15 December 2018).

63 *National Hazardscape Report*, Ministry of Civil Defence and Emergency Management, p.62, www.civildefence.govt.nz/assets/Uploads/publications/national-hazardscape-report-sept-2007-part3.pdf (accessed 20 February 2019).

64 Andrew McRae, 'A year after Edgecumbe floods: 78 homes still unoccupied', Radio New Zealand, 6 April 2018, www.radionz.co.nz/news/national/354321/a-year-after-edgecumbe-floods-78-homes-still-unoccupied (accessed 18 December 2018).

65 *Climate Change Implications for New Zealand*, p.5. See also Climate Change Adaptation Technical Working Group, *Adapting to Climate Change in New Zealand*, Ministry for the Environment, 2017, p.40, www.mfe.govt.nz/sites/default/files/media/adapting-to-climate-change-stocktake-tag-report-final.pdf (accessed 18 December 2018).

66 These are just the risks from a high-carbon world, let alone other environmental damage associated with growth. To take one of countless other problems, a recent study found that long-term exposure to unsafe air pollution could damage brains so much that people lose the equivalent of a whole year of schooling: see Xin Zhang, Xi Chen and Xiaobo Zhang, 'The impact of exposure to air pollution on cognitive performance', *Proceedings of the National Academy of Sciences*, 115, 37 (2018), pp.9193–97.

67 Nick Srnicek and Alex Williams, *Inventing the Future: Postcapitalism and a World Without Work*, Verso, London, 2016.

68 'The New Zealand Sectors Dashboard', Ministry of Business, Innovation and Employment, http://sectorsdashboard.mbie.govt.nz (accessed 18 December 2018).

69 Jo McKenzie-McLean, 'Robots to train vineyard and orchard workers to do a better job', Stuff, 11 October 2018, www.stuff.co.nz/technology/107776015/robots-to-train-vineyard-and-orchard-workers-to-do-a-better-job (accessed 5 December 2018).

70 'Quarterly Retail E-commerce Sales', U.S. Census Bureau News, 2018, www.census.gov/retail/mrts/www/data/pdf/ec_current.pdf (accessed 10 December 2018).

71 Steve Lohr, '"The Beginning of a Wave": A.I. Tiptoes into the Workplace', *New York Times*, 5 August 2018, www.nytimes.com/2018/08/05/technology/workplace-ai.html (accessed 18 December 2018).

72 One early example of this is the Tripartite Economic Alliance that has been announced between Auckland, Guangzhou and Los Angeles to

encourage economic collaboration between the three cities.

73 Simon Thomsen, 'Watch: Domino's just delivered its first pizza by drone', *Business Insider Australia*, 16 November 2016, www.businessinsider.com.au/dominos-just-became-the-first-business-to-deliver-food-by-drone-2016-11 (accessed 16 December 2018).

74 This would not be without costs but, as Chapter Two argued, the fewer the people who have to invest upfront in a technology relative to the number of people who will be impacted by it, the quicker the productive diffusion. Welcoming companies that quickly scale a technology, like Uber or Amazon, would be one of the fastest routes to adoption.

75 'Would you take a driverless bus?', Ericsson, www.ericsson.com/en/internet-of-things/trending/driverless-buses-in-stockholm-sweden (accessed 18 December 2018); Patrick Sisson, '9 cities with smart ideas to improve transportation', Curbed, 18 September 2018, www.curbed.com/2018/9/18/17873726/transportation-best-cities-bus-scooter-light-rail-public-transit (accessed 18 December 2018).

76 See Daron Acemoglu and Pascual Restrepo, 'Artificial Intelligence, Automation and Work', NBER Working Paper 24196, 2018, www.nber.org/papers/w24196.pdf (accessed 3 December 2018).

77 Moreover, their proposal has been challenged, and the specifics of any UBI are hard to get right and are contested in New Zealand and abroad. See Michael Fletcher, 'A Universal Basic Income may be a good idea – but we will still need social security that works', Briefing Papers, 30 March 2016, http://briefingpapers.co.nz/a-universal-basic-income-may-be-a-good-idea-but-we-will-still-need-social-security-that-works (accessed 18 December 2018).

78 'Historical Highest Marginal Income Tax Rates', Tax Policy Center, www.taxpolicycenter.org/statistics/historical-highest-marginal-income-tax-rates (accessed 16 February 2019).

79 See 'Table II.1. Statutory corporate income tax rate', OECD.Stat, https://stats.oecd.org/index.aspx?DataSetCode=TABLE_II1&_ga=2.265354381.323709462.1549144182-1474271205.1548444250 (accessed 2 February 2019). In 2018 the Trump administration passed a tax law that reduced the combined corporate tax rate to about 25 per cent. Clearly, however, the higher corporate tax rate previously had not stopped the rise of the numerous artificial-intelligence and automation companies, a sector in which the USA is a world leader.

80 Ilana Strauss, 'Would a Work-Free World Be So Bad?', *The Atlantic*, 28 June 2016, www.theatlantic.com/business/archive/2016/06/would-a-world-without-work-be-so-bad/488711 (accessed 18 December 2018).

81 See for example Tim Jackson, *Prosperity Without Growth*, Earthscan, Abingdon, 2009.

82 In New Zealand there are over a hundred men's sheds registered and in Australia nearly a thousand, with similar initiatives growing rapidly from the United Kingdom to Finland. 'Latest News', MenzShed, http://menzshed.org.nz/news/latest-news (accessed 18 December 2018); Mike Jenn, 'Men's sheds are more than buildings – they can be lifesavers', *Guardian*, 6 November 2013, www.theguardian.com/commentisfree/2013/nov/06/mens-sheds-building-lifesaver-community (accessed 18 December 2018).

83 'Shed heaven: The workshops turning men's lives around', BBC, 11 May 2017, www.bbc.com/news/uk-scotland-39871206 (accessed 18 December 2018).

84 Derek Thompson, 'A World Without Work', *The Atlantic*, July–August 2015, www.theatlantic.com/magazine/archive/2015/07/world-without-

work/395294 (accessed 18 December 2018).

85 This scheme is particular to the UAE's culture and society, and clearly may be viewed as problematic in other countries. See 'Marriage', Official Portal of the UAE Government, https://government.ae/en/information-and-services/social-affairs/marriage (accessed 18 December 2018).

86 David Grigg, 'Agriculture in the World Economy: An Historical Geography of Decline', *Geography*, 77, 3 (1992), pp.210–22.

87 *Review of Agricultural Policies in Korea*, OECD, Paris, 1999, p.29, www.oecd.org/korea/40417830.pdf (accessed 18 December 2018).

88 After a lot of efforts by unions and others, this was formalised for most workers by the Industrial Conciliation Amendment Act 1936 and the Factories Amendment Act 1936, which created a forty-hour week as the standard for most workers. See Peter Clayworth, 'Weekends – The great Kiwi weekend', Te Ara – the Encyclopedia of New Zealand, www.TeAra.govt.nz/en/weekends/page-3 (accessed 18 December 2018).

89 Jonathan Boston, *Safeguarding the Future: Governing in an Uncertain World*, Bridget Williams Books, Wellington, 2017.

Chapter Eight

1 See Maddy Savage, 'What really happened when Swedes tried six-hour days?', BBC, 8 February 2017, www.bbc.com/news/business-38843341 (accessed 17 December 2018); see calls by unions in the UK for shorter weeks in Sarah O'Connor, 'Trade unions seek role in age of automation', *Financial Times*, 20 November 2018, www.ft.com/content/6d50b080-ad56-11e8-8253-48106866cd8a (accessed 17 December 2018).

2 New Zealand has recently taken valuable steps in establishing ways to think about different paths for the country through the Treasury's Living Standards Framework, which aims to help decision-makers judge potential policies against more than just their economic results. Although my perspective here is slightly broader and reflective of my personal views, the living standards framework and focus on well-being is a good step to help us think clearly about what we might wish to prioritise as we shape the future of work.

3 As Chapter Six suggested, while most studies do not, a few studies do find evidence of a point at which more income no longer increases life satisfaction; see Ed Diener, Ronald Inglehart and Louis Tay, 'Theory and validity of life satisfaction scales', *Social Indicators Research*, 112, 3 (2013), pp.112–497, https://doi.org/10.1007/s11205-012-0076-y (accessed 16 January 2019). However, the level of income at the satiation point, US$125,000 for a household in Australia and New Zealand (if taken seriously as *the* number) is well above what the average New Zealand household earns and far above what poorer members of society earn. It is also likely to be difficult to reach through a UBI and gift stocks.

4 On cross-country comparisons of life satisfaction, see Betsey Stevenson and Justin Wolfers, 'Subjective Well-Being and Income: Is There Any Evidence of Satiation?', *American Economic Review*, 103, 3 (2013), pp.598–604; Betsey Stevenson and Justin Wolfers, 'Economic growth and subjective well-being: Reassessing the Easterlin Paradox', *Brookings Papers on Economic Activity*, 39, 1 (2008), pp.1–102. On health, see Chapter 2 of J. Helliwell, R. Layard and J. Sachs, *World Happiness Report 2017*, Sustainable Development Solutions Network, New York, 2017.

5 See A. Milner and A.D. LaMontagne, 'Underemployment and mental health: Comparing fixed-effects and random-effects regression

approaches in an Australian working population cohort', *Occupational and Environmental Medicine,* 74 (2017), pp.344–50; Maureen Dollard and Anthony Winefield, 'Mental health: Overemployment, underemployment, unemployment and healthy jobs', *Australian e-Journal for the Advancement of Mental Health*, 1, 3 (2002), pp.170–95, https://doi.org/10.5172/jamh.1.3.170 (accessed 18 December 2018); and Jenny Marlar, 'The Emotional Cost of Underemployment', Gallup, 9 March 2010, https://news.gallup.com/poll/126518/emotional-cost-underemployment.aspx (accessed 17 December 2018).

6 Jinmoo Heo, Youngkhill Lee, Paul M. Pedersen and Bryan P. McCormick, 'Flow Experience in the Daily Lives of Older Adults: An Analysis of the Interaction between Flow, Individual Differences, Serious Leisure, Location, and Social Context', *Canadian Journal on Aging/La Revue Canadienne du Vieillissement*, 29, 3 (2010), pp.411–23.

7 See 'Cognitive Diversity: AI & The Future of Work', Tata Communications, 2018, www.tatacommunications.com/wp-content/uploads/2018/09/Report_Cognitive-Diversity_AI-and-The-Future-of-Work.pdf (accessed 17 December 2018); and 'The Automation Advantage', Alpha Beta, 2017, www.alphabeta.com/wp-content/uploads/2017/08/The-Automation-Advantage.pdf (accessed 17 December 2018).

8 This is not an argument against any form of UBI in any context, but it is to suggest that, given the evidence today, I do not favour aiming for a UBI that is designed primarily to replace both work and work income in response to the threat of technological unemployment. The debate on a UBI in contexts in which the economy and society are still broadly based on work is a wider one and not the focus of this book. It is worth noting that the benefits and challenges in those contexts are different. For example, one advantage could be that a UBI would help create lower effective marginal tax rates, and therefore encourage people not working to return to work, as they would not lose their benefit when they did so. Other potential advantages would be to reduce the negative attitudes towards people on benefits from those still working by making benefit payments universal, and to provide recognition for domestic and childcare work that often goes unpaid.

9 This example draws from Robert Pollin, 'De-growth vs a green new deal', *New Left Review*, 112 (July–August 2018), pp.5–25.

10 This is over the 1995 to 2011 period. Denmark and Germany are not the only countries to achieve this but they have achieved it by far the most clearly and significantly, suggesting their achievement is likely to be robust regardless of the methodology for measuring carbon decoupling. See OECD, *Green Growth Indicators 2017*, OECD Green Growth Studies, OECD Publishing, Paris, 2017, p.35.

11 'Few countries are pricing carbon high enough to meet climate targets', OECD, 18 September 2018, www.oecd.org/newsroom/few-countries-are-pricing-carbon-high-enough-to-meet-climate-targets.htm (accessed 17 December 2018).

12 This phrase is widely attributed to the economist John Maynard Keynes.

Acknowledgements

The fact this book has happened is a testament to the great help and support I have received from others along the way.

Tom Rennie at Bridget Williams Books strongly encouraged me to consider writing about the future of work in New Zealand. He has been a patient guide, supporter and advisor and has replied to countless queries from me with unfailing good humour. The broader team at Bridget Williams books – Julia Wells, Nancy Swarbrick and Sasha Francis – has been equally helpful and patient in arranging peer reviewers, readying the book for publication, and organising events and publicity. Caren Wilton superbly edited the manuscript and Barbara Graham expertly guided the book through the final stages of production. Jo Bailey provided a wide range of cover designs and happily kept adjusting to get the result we were looking for. I am also very grateful to Bridget Williams who was willing to believe in this project and also provided helpful comments and advice along the way.

In writing this book I've benefited greatly from speaking with a range of people who generously gave up their time, including David Brougham, Stuart Trundle, Max Rashbrooke, Phil O'Reilly, Nick Hill, Patrick McVeigh, Jane Parker, Paula O'Kane, Sara Walton, Diane Ruwhiu, Kevin Jenkins, Jonathan Boston, David Autor, Daron Acemoglu, Lant Pritchett, Ricardo Hausmann, Zeynep Ton, James Bessen, Lawrence Summers, Diego Comin, Robert Lawrence, Charles Kenny and an anonymous New Zealand civil servant.

I also received valuable feedback on the original outline and chapters from Hannes Malmberg, Gail Pacheco, Hayden Glass, John Holt, Bill Rosenberg, Nicholas Agar, Grace Millar, Jim McAloon, Simon Chapple, Stella Chrysostomou, and two anonymous reviewers. I am particularly grateful to Paul Conway who provided detailed comments on every chapter and to Max Harris for his comments on the chapters as well as thoughtful advice on multiple occasions.

In writing the book my thinking has been sharpened and spirits bolstered by conversations with friends, including Jennifer Austin, Daniel Garrote Sanchez, Matthew Collin, Duncan Kauffman, Kartika Bhatia, Cindy Paladines, Samik Adhikari, Hautahi Kingi, Yashodhan Ghorpade, Fraser Guillen and Tom Scrase.

Finally, I would like to thank my mother, Gwenny, who has provided unstinting moral and intellectual support as well as comments on various chapters. I've benefited from numerous discussions about New Zealand's economy and environment with my father, Guy, who also generously provided comments on all the chapters. My brother, Erin, has helped keep my spirits and motivation up and regularly provided a breath of fresh perspective. Guy's wife, Catharine, provided helpful comments on the cover designs in particular. My partner, Lucila, has provided moral support and encouragement, suggested excellent new ideas, and read and improved chapters. She even helped with fiddly endnotes as deadlines loomed and has shown enormous patience as yet another evening or weekend saw me either disappear to work on the book (or disappear into my head thinking about it while I was meant to be doing something else!)

The errors that remain are, of course, all mine.

Index